SPECTROSCOPIC ANALYSIS OF METAL COMPLEXES

BY
IFFAT IMTIAZ

NORTHWEST
University Press

This book is printed on acid-free paper.

Published by NorthWest University Press.

Printed in the United States of America
ISBN: 978-0-6151-4832-8

This book is printed on 6" x 9", perfect binding, 60# cream interior paper, black and white interior ink, 100# white exterior paper, full-color exterior ink. Prices are subject to change.

Cover Title Designed by NorthWest University Press.

Spectroscopic Analysis of Metal Complexes
First Edition
Iffat Imtiaz

"To my parents, brothers, sister and all those,
who enlightened me when I walked on the road,
which goes towards the mines of gold."

- Iffat Imtiaz -

ACKNOWLEDGEMENT

All praise is for ***Allah Almighty***, the most bountiful, omnipotent and merciful who created this world of unprecedented actions and gifted humans with the intellect of thinking, performing and selected them to explore the hidden secrets of his titanic kingdom. My all admirations are for the most commendable ***Prophet Muhammad*** (May peace be upon Him) who by His sayings and doings showed me the right path to follow i.e. Islam.

Words are wane in expressing my veneration for my parents, who sacrificed their desires over mine, prayed for my edification, endeavored for my betterment, faced with courage all vexatious moments that came along and cheerfully took burden of my prodigious and prodigal studies. Words are inadequate to express my heartiest adoration for my brothers who embosomed me with the everlasting shadows of love and affection and emboldened me in removing the impediments that came in my way during this strenuous work.

I am enormously indebted to ***Prof. Dr. M. Mazhar***, Chairman, Department of Chemistry, Quaid-i-Azam University (Pakistan), for providing access to all facilities required for this project.

Thanks are due to ***Dr. Amin Badshah***, my supervisor, for his guidance during this work. I ought to submit my thanks to all teachers of Inorganic/Analytical Chemistry Section especially ***Prof. Dr. Saqib Ali, Dr. Mrs. Zareen Akhtar, Dr. S.A. Tirmiz*** and also ***Syed Munir Hussain Shah***. Their sympathetic attitude always encouraged me to solve my research problems.

Thanks are due to ***Dr. Bushra Mirza***, Department of Biological Sciences, Quaid-i-Azam University, Pakistan, for antibacterial activity assignment and ***Sir Sajjad Ahmed*** for ICP-ES analysis.

Acknowledgements are also due to my bosom friends and colleagues, the teaching staff, the official and technical staff of Chemistry Department for their kindly cooperation, valuable suggestions and encouragement throughout this work.

ABSTRACT

In this study, the new metal complexes of sulfamethoxazole (antibiotic) with palladium (II), iron (II), iridium (III) and tin (II) cations have been synthesized and characterized using infrared and NMR.

The infrared spectra of the samples were recorded in the range 4000–400 cm^{-1} and their fundamental vibrational wave numbers were obtained. The vibrational assignments were determined by using the group frequency tables and compared with the wave numbers of SMX found in the literature. Studying the infrared spectra of the metal complexes reveals that the active binding sites of SMX are the isoxazole ring nitrogen and sulfonic oxygen. The proposed structure for metal complexes of SMX derived from NMR spectra is in consistent with the infrared results. ICP-ES, AAS and UV-visible were used for determining the percentage of metal in metal complexes.

Metal complexes of pyrimethamine (antiparasitic) with zinc (II), cobalt (II) and cadmium (II) cations have also been synthesized. These complexes have been characterized on the basis of FT-IR, ^{1}H-NMR, ^{13}C-NMR, UV-visible spectrophotometer, AAS and ICP-ES. From FT-IR results it is concluded that nitrogen of secondary amine and nitrogen of conjugated ring act as binding sites with metal ion. Structure elucidated by NMR for pyrimethamine metal complexes are in consistent with the FT-IR results. Antibacterial activity of these complexes has also been tested and compared with that of parent drugs (pyrimethamine,sulfamethoxazole) and standard drugs.

It is found that antibacterial activity of pyrimethamine and sulfamethoxazole is increased on coordination with metals. Biological activity of the complexes shows that compound iridium sulfamethoxazole and tin sulfamethoxazole shows the significant antibacterial activity as compared to parent drug and standard drug against *Pseudomonas picketii, Bordetella bronchiseptica, Escherichia coli* and good antibacterial activity against *Micrococcus luteus.* Biological activity of pyrimethamine shows that they show low antibacterial activity against *Bordetella bronchiseptica* and *Escherichia coli* as compared to standard drug Cefixime and Roxyithromycin, but good antibacterial activity as compared to parent drug.

TABLE OF CONTENTS

SPECTROSCOPIC ANALYSIS OF METAL COMPLEXES

INTRODUCTION

Since late 19th century extensive efforts have been made to develop coordination chemistry, which have remained successful with the discovery and characterization of thousands of coordination complexes, their research is still on the way for synthesis of complexes which can be used in biological system. Metal binding substances, many of which function by chelation, form a class of substances which have furnished many useful drugs and other substances of value in selective toxicity [1].

1.1 Theoretical Background

1.1.1 Coordination Compounds

A coordination compound consists of a central metal atom or ion surrounded by a cluster of atoms, ions or molecules . The complex tends to retain its identity even in solution, although partial dissociation may occur.It may be a cation or an anoin or nonionic, depending on the sum of the charges of the central metal atom and the surrounding ions and molecules [2].

In the complex formation each molecule or atom donates a pair of electrons to the empty orbitals of central atom, thus forming a semi-covalent bond, for example, in hexamine cobalt(III) complex, each of the ammonia group donates a lone pair of electron to cobalt(III) cation, resulting in formation a complex ion.

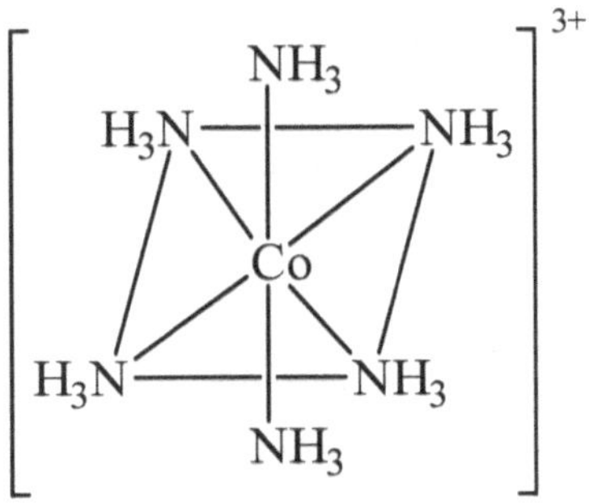

The total charge on the complex ion is the resultant of the charge on the metal ion and the coordinated groups. The total number of bonds which the ligand forms with the central atom in a complex is called its "coordination number". Coordination number varies from 2 to 12 of which 4 and 6 are more common and important. Complex compounds with coordination number four have five principal symmetries, i.e., Td, D_{4h}, C_{4v}, C_{3v} and C_{2v} and those with coordination number six have Oh and C_{4v} symmetry.

1.1.2 Stability of Metal Complexes

The stability of complexes obviously depends on the nature of metal and that of the ligand.

With reference to metal the following factors are important that influence the stability of metal complex.

i) Size and Charge [3]

The smaller the size and the larger the charge of the metal ion, the more stable are the metal complexes. Thus stability of favoured by a large charge-to-radius ratio of the metal ion. The stabilities of high spin metal complexes of the M^{+2} ions between Mn^{+2} and Zn^{+2} with a given ligand frequently vary in the order $Mn^{+2} < Fe^{+2} < Co^{+2} < Ni^{+2} < Cku^{+2} > Zn^{+2}$. This is called 'natura. order of stability'

ii) Crystal Field Effects [3]

Depending upon the nature of ligand, the metal d orbitals split into different energies of different field (e.g., octahedral, tetrahedral etc.). In octahedral field the d orbitals split into t_{2g}, which is lower in energy, and e_g, which is higher in energy, sets of orbitals. When an electron is placed in lower energy orbitals, energy is lost by the system and the complex is stabilized by $4D_q$ energy. As the number of electrons residing in the t_{2g} oribitals increase, stabilization also increases. If an electron is placed in e_g set, complex will be destabilized by $6D_q$ and its stability is decreased. Therefore, one can determine the stability order on the basis of crystal field effects for metal ions of 3-d series. From the crystal considerations, high spin complexes of metal ions with d^o, d^5 and d^{10} configurations should be of minimum stability and those with d^3 and d^8 configuration should have maximum stability while stability of other complex ions should be in between these two extremes.

iii) Hard and Soft Acids and Bases [3]

All metal ions are grouped as acids and ligands as bases. Depending on the nature of metals and ligands, they are further grouped as hard or soft acids and bases. Metal ions which have small radii, carrying a high charge and do not have easily distortable valence shell electrons (i.e., less polarizable) are termed hard acids. Class 'a' metals (e.g., Na^{+}, Al^{+++}, the lanthanides, Ti^{++++} and Fe^{+++}) are included in hard acids. The soft acid group includes metals which are large, have low charge and their valence shell electrons can be easily polarized. Most of the class 'b' metals (e.g., Pt^{++}, Pd^{++}, Hg^{++}, Pb^{++}, Rh^{++}) are termed as soft acids. Similarly, ligands have been divided into two groups i.e., hard bases, which are non-polarizable and soft bases, which are easily polarizable. In this group, ligand containing small sized donor atoms such as O, N, F (difficult to polarize) are termed as hard bases while those with larger donor atoms (like S, Se, P, As, Br, I, etc.) are easily polarizable and are called as soft bases. There are certain metal ions and ligands which have the characteristics of both groups and are termed as border line cases.

Soft acids interact strongly with soft bases and vice versa. Therefore, complexes formed by the interaction of hard acids (metal ions) with hard bases are more stable than formed by hard acids and soft bases. Most of the first row transition metal ions are hard acids and form stable complexes with hard base ligands containing O, N, F door atoms. Complexes such as $[Cr(OH_2)_6]^{+3}$, $[Co(NH_3)_6]^{+3}$, $[Cr(NCS)_6]^{-3}$ are expected to be more stable than $[Cr(SR_2)_6]^{+3}$, $[Co(PPh_3)_6]^{+3}$, $[Cr(SCN)_6]^{-3}$ ions. With reference to the role of ligand in determining the stability of metal complexes the following factors are important.

i) Base Strength [3]

The base strength of a molecule is a measure of the stability of the complex that the molecule forms with H^+ ion. The greater the base strength of a ligand, the greater is the tendency of the ligand to form stable complexes with class 'a' metals. It is reasonable that ligands which bind H^+ firmly should also form stable complexes with class 'a' metal ions. From this point of view F should form more stable complexes than Cl^-, Br^- or I^-.

ii) Chelate Effect [3]

Complexes of ligands which are capable of forming chelates are more stable as compared to those which occupy only single coordination position. Complexes such as $[Ni(en)_3]^{+2}$ or $[Ni(dien)_2]^{+2}$ are much more stable than $[Ni(NH_3)_6]^{+2}$. Stability of metal chelates increases with number of donor atoms in a ligand. A tridentate ligand which is capable of coordination at three positions, (e.g., ethylene triamine) forms more stable complex with a metal ion than a bidentate ligand (e.g., ethylene-diamine) which is capable of coordination at two position. Similarly, a hexadentate ligand (e.g., EDTA) which coordinates at six positions of a meal ion, forms most stable complexes. The increased stability due to chelation originates from increase of entropy and favourable changes in enthalpy.

iii) Chelate Ring Size [3]

Stability of chelates depends upon the size of ring as well. There are examples of complexes containing chelates of varying number of atoms such as 4, 5, 6, 7, . . . but their stabilities widely differ. Usually five membered chelate rings without any π-bonds and six membered chelate with a delocalized π-bond system are most stable. Four membered chelates formed by polynuclear complexes and also anionic ligands such as carbonates, sulphates, etc. are very unstable. It is noted that Co(III) complexes of 1,2-diaminopropane are more stable than those of 1,3-diaminopropane which form a saturated six membered chelate. Similarly, six membered chelates of anion acetylacetonate are stabilized due to delocalized π-electrons in it.

iv) Steric Strain [3]

Stability of complexes also depends upon the geometry and type of substituents in organic ligands. This is specially true for complexes of multidentate ligands in which donor atoms are present at certain angles hindering coordinate bond formation. Large bulky ligands form less stable metal complexes than do analogous smaller ligands, e.g., $H_2NCH_2CH_2NH_2$ forms more stable complexes than $(CH_3)_2NCH_2CH_2N(CH_3)_2$.

The consequences of charge, size, CF stabilization, etc. on the stability of metal complexes are vitally important to the chemistry of coordination compounds. For example, the oxidation potentials for metal ions change markedly as the type of ligand is changed (Table 1.1). As water is replaced by CN^-,EDTA, or NH_3 in complexes of Fe^{2+} and Co^{2+}, the tendency for oxidation to the M^{3+} state is markedly enhanced. These ligands form very much more stable complexes with M^{3+} ions than with M^{2+} ions; this provides the driving force for the oxidation. This is particularly true in Co^{2+} systems. The complex $[Co(H_2O)_6]^{3+}$ will oxidize H_2O to O_2; in contrast, aqueous solutions of Co^{2+} salts are readily oxidized by atmospheric O_2 to $[Co(III)L_6]$ complexes in the presence of ligands such as NH_3, CN^-, NO_2^-. The large change in oxidation potential that results from the presence of these ligands is largely due to the effect that the ligands provide agreater CF than water provides; this favours the conversion of high-spin d^7 Co^{2+} compexes to the highly CF- stabilized low-spin Co^{3+} d^6 complexes.

Table 1.1: The Oxidation Potentials for some Cobalt and Iron Complexes

Reaction	Oxidation Potential
$[Fe(H_2O)_6]^{2+} \rightarrow Fe(H_2O)_6]^{3+} + e$	-0.77 V
$[Fe(CN)_6]^{4-} \rightarrow [Fe(CN)_6]^{3-} + e$	-0.36 V
$[FeEDTA]^{2-} \rightarrow [FeEDTA]^- + e$	+0.12 V
$[Co(H_2O)_6]^{2+} \rightarrow [Co(H_2O)_6]^{3+} + e$	-1.84 V
$[Co(NH_3)_6]^{2+} \rightarrow [Co(NH_3)_6]^{3+} + e$	-0.10 V

1.1.3 Coordination compounds in Biological system

It has been proved by recent research findings that many of the normal life processes in the biological system involve coordination compound reactions. For example iron is present in the human body in the form of a complex compound (haemoglobin, myoglobin). It may have iron in oxidation state (II) or (III) depending upon complexing agent coordinated to it. It act as bivalent in haemoglobin and myoglobin and behaves as trivalent for catalases and oxidases. Iron has ability to form low and high spin complexes with the same oxidation state for example iron (II) yields high spin complexes with the haemoglobin and low spin complexes with the oxyhaemoglobin. Cobalt complexes are also present in biological system. Vitamin B_{12} is a cobalt (III) complex and some complexes of cobalt (II) with histadine and glycine act as oxygen carriers in the body [4]. Cobalt (II) is also found attached to low symmetry sites in enzymes. Copper with oxidation state (I) and (II) in haemocyanin also act as oxygen carrier. Molybdenum, the heaviest essential metal in the human body, occurs in oxidation state(V) and (VI) and even sometime (III) and (VI) oxidation states are involved [5]. Its role in xanthine and purine oxidases in milk is well understood in the light of coordination chemistry.

It has been observed that many metal chelates when present in body fluids or tissues are quite active on a wide range of microorganisms including even those which show high resistance to current antibiotics. For example metal chelates of 1, 10-phenanthroline exerts much more bacteriostatic effect as compared to the ligand itself in rumen bacteria , on acid fast bacteria and on number of gram positive microorganism [6].The metal chelates of sulfa drugs have been found to be more bacteriostatic than the drugs themselves [7 and 8].

Similarly, 6-Mercaptopurine (6MPH) is a biological important anti-carcinogen and the Pd (II) complex Na_2 [Pd (6MP) $_2Cl_2$].H_2O has been found to be an effective anti-tumor agent [9]. The anti-tumor activity of 6-methylmercaptopurine (6MMP) is about 20% that of 6-mercaptopurine and it is possible that the methyl derivatives is converted to 6-mercaptopurine in the host tissue [10]. Since it was thought that the S-demethylation in vivo might be accomplished by the agency of (b) class metal ions, an attempt was made to S-demethyllate $PdCl_2$ (bMMP), but only products of indefinite composition were isolated [11].

Thiouracil (thuH) has been used as a drug for the treatment of hyperthiroidism, since it inhibits the production of thyrixine. The Pd (II) complex $Pd(thu)_2$ is reddish brown. Its insolubility together with infrared data, suggest that the complex is polymeric with thiolo-bridges and the ligand is coordinated through sulphur only.

"Staniform" containing methyl stannic iodide was used as an external treatment drug for staphylococcal infection until 1958. "Stannoxyl" second drug which contain tin powder and tin(II) oxide in the form of tablets for oral treatment against disorder and abnormalities such as ache, boil, carbundes and styes was used until early 1980's [12]. Tin(II) fluoride is superior to all other fluoride for the inhibition of dental plaque formation [13]. Tin-haem or dichloro(protoporpharin IX)tin as shown in Fig.1.2 is a potent inhibitor of haemoxidase [14-18].

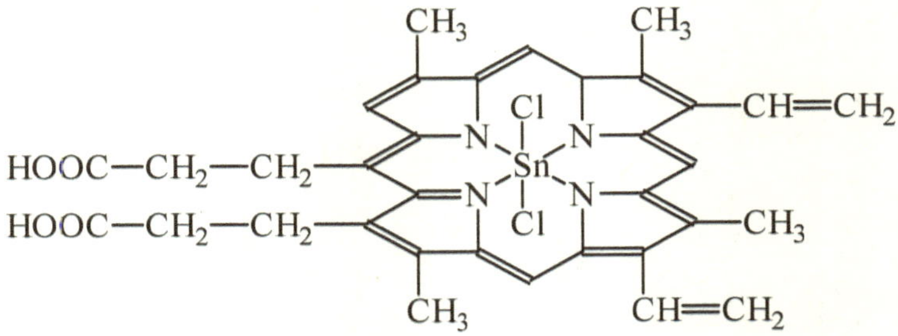

Figure 1.1

The first report organotin compounds possessing same antitumour activity were organotinoxides (R_2SnO), diorganotin hydroxides [$R_2Sn(OH)X$], distanoxanes [$(XR_2Sn)_2O$]$_2$ and di(methylcarbonylmethoxide) [$R_2Sn(CH_2COMe)_2$], where R = alkyl and X = halogen [19,20]. Brown demonstrated in 1972 that the growth of malignant tumours in mice was appreciably retarded by feeding the animal with triphenyltin acetate. The injection of this compound into blood stream had the same effect [21].

Complexes of tin(IV) with biological important molecules were found to have positive activity against tumour. Adenine and glycylglycine, a range of amino- acids, 2-mercaptoethanesulphate and purine-6-thiol are all ligands forming such complexes [22-23].

Organotin derivatives of donor ligand, particularly organotin carboxylates show promsing in vivo or in vitro antitumour activity has been described in various reports [24-27].

1.1.4 Effect of Metal Complexation on Biological Activity

Drugs play a vital role as bioligands in biological systems. Nitrogen containing bases such as pyrrole, pyridine, purine, pyrimidine, amines such as histamines, and carbohydrates such as pentose, glucose, and different vitamins such as Vitamin A are well known examples of bioligands.

In drugs if the biometals are incorporated, the activity of the metal is attained through the formation of complexes with different bioligands. Thermodynamic and kinetic properties of complexes govern the mode of the biological action. As the biological membranes are lipoprotein in nature therefore it is necessary to increase the lipophilicity of drug to increase the effective penetration to pass through biological membranes. Aspirin, thiosemicarbazone, anticancer drugs and antibiotics exhibit the drug action through complexsation with available biometals *in vivo* [28]. A number of diseases and their remedies are dependent upon the metabolism of inorganic constituents. Copper metabolism is somehow related with the rheumatoid arthritis. Chromium metal has an important role to play in case of diabetes.

A virus can penetrate into the host's cell only when it is mediated by some suitable metal ions. In virus replication process the role of metal ions is thus very important [29].For example, zinc present in the cell wall of the bacterium E. coli can coordinate in the sulphur site present in the virus coating. The penetration of viral DNA into the host cell can be arrested by using any suitable metal complex like $Cd(CN)_3$, which can preferably bind with the 'S' site as Virus-(S)-$Cd(CN)_3$. Thus by increasing or decreasing the concentration of copper and zinc, viral growth can be controlled because they are essentially required for virus replication.

It is anticipated that stable metal chelate exerts its effect by structural activation or in-activation of susceptible biological site and the response depends on the rate of association of chelate with susceptible site or its rate of dissociation from this site [30]. Maximum biological activity is associated with complex cations which are highly soluble and penetrate in biological membranes and at the same time have high capacity to bind the diverse biological surface by coulombic and vander waals forces [31].

In many cases the pharmacological activity of antibiotics after complexation with metals is enhanced as compared to free ligands [32] although in some cases the activity of antibiotics may decreases after metal complexation. For example, many of the well known antibiotics, penicillin, streptomycin, bacitracin, tetracycline, etc. act as chelating agents and their action is improved by presence of small amount of metal ions [33]. Polydentate ligands (having more than one lone pair or negative charge to form coordinate covalent bond with central metal ion) when binds with metal ion to form ring system is called chelation. In chelates chelates, metal is firmly held by a number of ligand atoms usually nitrogen oxygen or sulphur through coordinate covalent bond. Some chelates are model analogues of certain metallo-enzymes [34]. Furthermore some chelates develop considerable fungicidal and antimicrobial activity [35].

1.2 Ligands Used

Ligands used for synthesis of metal complexes in this research work are

i) Sulfamethoxazole

ii) Pyrimethamine

1.2.1 Pyrimethamine

Chemical name:	5-(4-Chlorophenyl)-6-ethyl-2,4-pyrimidinediamine
Chemical formula:	$C_{12}H_{13}ClN_4$
Molecular weight:	248.71 g/mol

Structure:

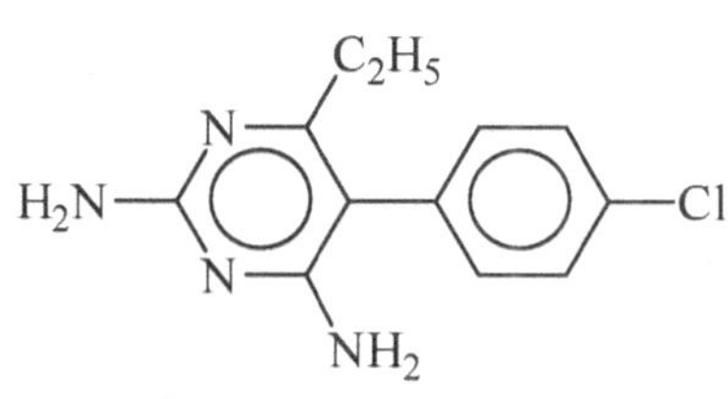

Pyrimethamine is an atiparasitic compound available in table form for oral administration.

1.2.2 Sulfamethoxazole

Chemical name:	4-Amino-N-(5-methyl-3-isoxazolyl)-benzenesulfonamide
Chemical formula:	$C_{10}H_{11}N_3O_3S$
Molecular weight:	253.28 g/mol.

Structure:

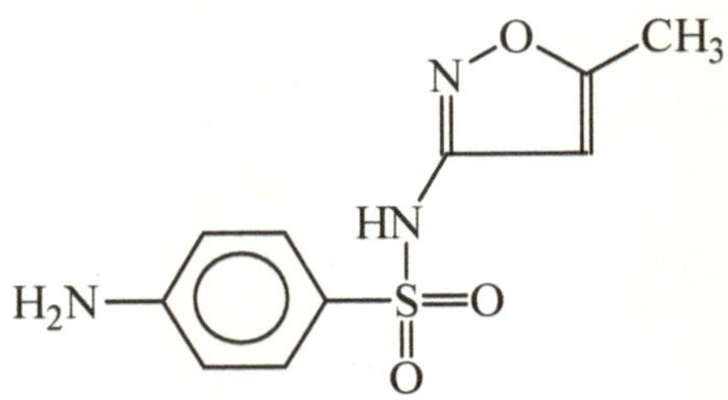

Physical state:	White crystalline powder, colourless and tasteless
Stability:	Stable under ordinary conditions

General Description and Application

Sulfonamide is an organic sulfur compounds containing the radial $-SO_2NH_2$, an amine moiety ($-NH_2$) and a benzene ring [36-38]. Its molecular structure is similar to p-aminobenzoic acid (PABA) which is needed in bacteria organisms as a substrate of the enzyme dihydropteroate synthetase for the synthesis of tetrahydrofolic acid (THF).

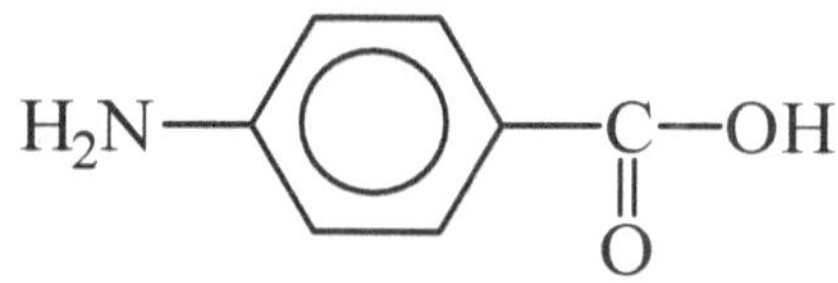

Para-aminobenzoic acid (PABA)

A final requirement for sulfonamides is that the amine group possesses a pk_a of 4.7 and the sulfonamide moiety possesses a pk_a of 6.5 [36]. The basis for these requirements is to mimic the structure of PABA and subsequently produce antibacterial effects.

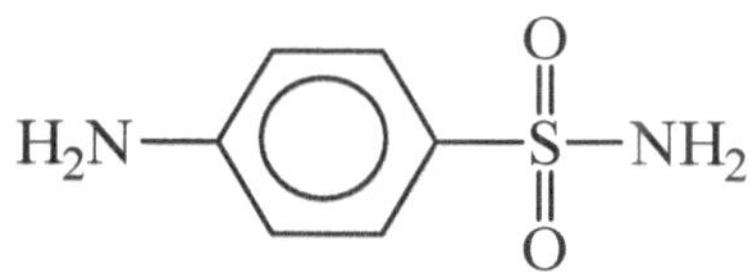

Sulfanilamide

Mono-substitution on the sulfonamide group is the basis for pharmacokinetic variation within this class of antibacterial. The addition of different functional groups enables alteration and manipulation of the properties of the drug. For instance, both sulfamethoxazole and sulfadoxine have a pk_a of 6.1, however, the half-life of sulfamethoxazole is 11 hours whereas the half-life of sulfadoxine is 150 hours [36]. It appears that functional groups that impart beneficial antibacterial effects are those that lower the sulfonamide pk_a, specifically electron withdrawing groups (EWGs) and heterocyclic rings [36]. Accordingly, the addition of various functional groups allows medicinal chemists to manipulate pharmacokinetic properties of the drug and alter the properties of sulfonamides.

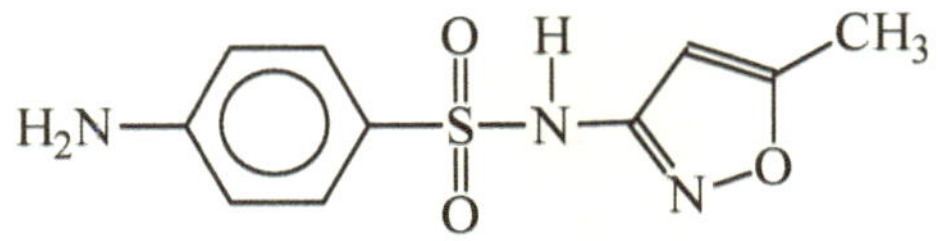

Sulfamethoxazole

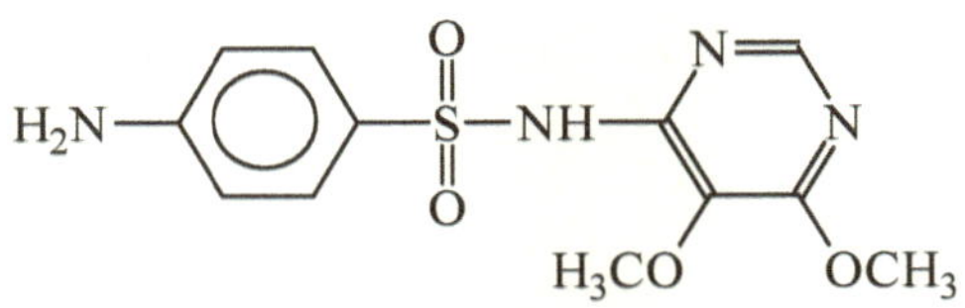

Sulfadoxine

Sulfonamides, derived from chiefly sulfanilamide are capable of interfering with the metabolic processes in bacteria that require PABA. They act as antimicrobial agents by inhibiting bacterial growth and activity and called sulfa drugs. They are used in the prevention and treatment of bacterial infections, diabetes mellitus, edema, hypertension and gout [39].

Sulfamethoxazole is used particularly for the prophylaxis and treatment of acute urinary tract infections and of pyodermata and infections of wounds and soft tissues [40]. It is usually used in combination with trimethoprim, a bacterial dihydrofolate reductase inhibitors sulfonamide drug in the ratio 5:1 [41]. The trade names of the combined product are Bactrim and Septra.

Microbiology [42]

Sulfamethoxazole inhibits bacterial synthesis of dihydrofolic acid by competing with *para*-aminobenzoic acid (PABA). Trimethoprim blocks the production of tetrahydrofolic acid from dihydrofolic acid by binding to and reversibly inhibiting the required enzyme, dihydrofolate reductase. Thus, Sulfamethoxazole and Trimethoprim Oral Suspension blocks two consecutive steps in the biosynthesis of nucleic acids and proteins essential to many bacteria.

In vitro studies have shown that bacterial resistance develops more slowly with Sulfamethoxazole and Trimethoprim Oral Suspension than with either sulfamethoxazole or trimethoprim alone.

In vitro serial dilution tests have shown that the spectrum of antibacterial activity of Sulfamethoxazole and Trimethoprim Oral Suspension includes the common urinarytract pathogens with the exception of Pseudomonas aeruginosa. The following organisms are usually susceptible: Escherichia coli, Klebsiella species, Enterobacterspecies, Morganella morganii, Proteus mirabilis and irtdole-positive Proteusspecies including Proteus vuigaris.

The usual spectrum of antimicrobial activity of Sulfamethoxazole and Trimethoprim Oral Suspension includes the following bacterial pathogens isolated from middle ear exudates and from bronchial secretions: Haemophilus influenzae, including ampicillin-reststant strains, and Streptococcus pneumoniae. Shigella flexneriand Shigella sonnei are usually susceptible. The usual spectrum also includes enterotoxigenic strains of Escherichia coli (ETEC) causing bacterial gastroenteritis.

REPRESENTATIVE MINIMUM INHIBITORY CONCENTRATION VALUES FOR ORGANISMS SUSCEPTIBLE TO SEPTRA (MIC-μg/mL) [42]

Bacteria	TMP alone	SMX alone	TMP/ SMX (1: 20)	
			TMP	SMX
Escherichia coli	0.05 - 1 .5	1.0 - 245	0.05- 0.5	0.95 - 9.5
Escherichia coli (enterotoxigenic strains)	0.015 - 0.15	0.285 - > 950	0.005 - 0.15	0.095 - 2.85
Proteus species (indole positive)	0.5 - 5.0	7.35 - 300	0.05 - 1.5	0.95 - 28.5
Morganella morganii	0.5 - 5.0	7.35 - 300	0.05 - 1 .5	0.95 - 28.5
Proteus mirabilis	0.5 - 1.5	7.35 - 300	0.05 - 0.15	0.95 - 28.5
Klebsiella species	0.15 - 5.0	2.45 - 245	0.05 - 1 .5	0.95 - 28.5
Enterobacter species	0.15 - 5.0	2.45 - 245	0.05 - 1 .5	0.95 - 28.5
Haemophilus influenzae	0.15 - 1 .5	2.85 - 95	0.015 - 0.15	0.285 - 2.85
Streptococcus pneumoniae	0.15 - 1 .5	77.35 - 24.5	0.05 - 0.15	0.95 - 2.85
Shigella flexneri *	<0.01 - 0.04	< 0.16 - > 320	< 0.002 - 0.03	0.04 - 0.625

TMP= trimethoprim; **SMX**= sulfamethoxazole
*Rudoy RC, Nelson JO, Haltalin KC. ***Antimicrobial Agents And Chemothel:*** May 1974; 51439- 443.

1.3 Characterization of Synthesized Complexes

Structural elucidation of new organic, polymeric, and organometallic compounds can be facilitated utilizing spectroscopic techniques such as infrared (IR), nuclear magnetic resonance (NMR), and mass spectrometry (MS). Qualitative identification of functional groups, structural conformation and molecular weight are determined for unknowns and molecular structure are assigned. In addition for materials that are mixtures,it is usually possible to deconvolute the spectra in order to identify and quantitate the components.

UV-visible spectroscopy, Atomic Absorption spectroscopy (AAS), inductively coupled plasma atomic emission spectrometer (ICP-ES) were used to determine the percentage of metal in metal complexes.

1.3.1 The Infrared Spectrum [43]

Of all the properties of an organic compound, the one that, by itself, gives the most information about the compound's structure is its infrared spectrum. A molecule is constantly vibrating: its bonds stretch (and contract), and bend with respect to each other. Changes in vibrations of a molecule are caused by absorption of infrared light: light lying beyond (lower frequency, longer wavelength, less energy) the red of the visible spectrum. A particular part of the infrared spectrum is referred to either by its wavelength or – and this is considered preferable – by its frequency. Wavelength is expressed in microns, μm (1 μm = 10^{-4} cm or 10^4 Å). Frequency is expressed, not in hertz, but in wavenumbers, cm^{-1}, often called reciprocal centimeters; the wavenumber is simply the number of waves per centimeter, and is equal to the reciprocal of the wavelength in centimeters.

Infrared spectrum helps to reveal the structure of a new compound by telling us what groups are present in – or absent from – the molecule. A particular group of atoms gives rise to characteristic absorption bonds; that is to say, a particular group absorbs light of certain frequencies that are much the same from compound to compound.

Table 1.2 showing the absorption frequency of various functions groups (cm−1). Interpretation of an infrared spectrum is not a simple. Overtones (harmonics) may appear at just twice the frequency of the fundamental band. The absorption band of a particular group may be shifted by various structural features – conjugation, electron withdrawal by a neighboring substituent, angle strain or Van der Waals strain, hydrogen bonding and be mistaken for a band of an entirely different group.

Table 1.2: Characteristic infrared absorption frequencies of various functional groups (cm^{-1}).

Group	Band	Remarks	Ref
Free−OH	3650-3590 (v)	Sharp; OH stretching	[44]
$-NH_2$	3400-3100 (m)	Stretching	[45]
C−H	3000-3100 (m)	Aromatic ring	[43]
$\rangle CH_2$, $-CH_3$	2960-2850 (s)	Two or three bands	[45]
⟩C=N− C−N (amines)	1690-1640 (v) 1180-1360 (s)		[44]
Conjugated cyclic system	1660-1480 (v)		[44]

s = strong; m = moderate; v = variable

1.3.2 The Nuclear-Magnetic Resonance (NMR) Spectrum [43]

Like electrons, the nucleus of certain atoms is considered to spin. The spinning of these charged particles – the circulation of charge-generates and "Magnetic Moment" along the axis of spin, so that these nuclei act like tiny bar magnets. One such nucleus and the one we shall be mostly concerned with is the proton, the nucleus of ordinary hydrogen, ^1H.

Now, if a proton is placed in an external magnetic field, its magnetic moment, according to quantum mechanics, can be aligned in either of two ways: with or against the external field. Alignment with the field is more stable, and energy must be absorbed to "flip" the tiny proton magnet over to the less stable alignment against the field.

Just how much energy is needed to flip the proton over depends, as we might expect, on the strength of the external field: the stronger the field, the greater the tendency to remain lined up with it, and the higher the frequency (Remember $\Delta\varepsilon = h\nu$) if the radiation needed to do the job.

$$\gamma = \frac{\gamma H_o}{2\pi}$$

Where ν = frequency in Hz

H_o = strength of the magnetic field, in gauss

γ = a nuclear constant, the magnetogyric ratio, 36, 750 for the proton

In a field of 14, 092 gauss, for example, the energy required corresponds to electromagnetic radiation of frequency 60 MHz (60 mega hertz or 60 million cycles per second): radiation in the radio frequency range, and of much lower energy (lower frequency, longer wavelength) than even infrared light.

In principle, we could place a substance in a magnetic field of constant strength, and then obtain and spectrum in the same way was obtain an infrared or an ultraviolet spectrum: pass radiation of steadily changing frequency through the substance, and observe the frequency at which radiation is absorbed. In practice, however, it has frequencies found for most carbon atoms in organic structure. The scale shows the common range of values, which are always written from right to left, a convention that denotes frequencies at higher and positive on the left and lower and positive on the right. The TMS signal, would be on the extreme light on zero, by definition. The frequency end of the spectrum, on the left with high values, is described as being downfield, and right side with low values is said to be upfield.

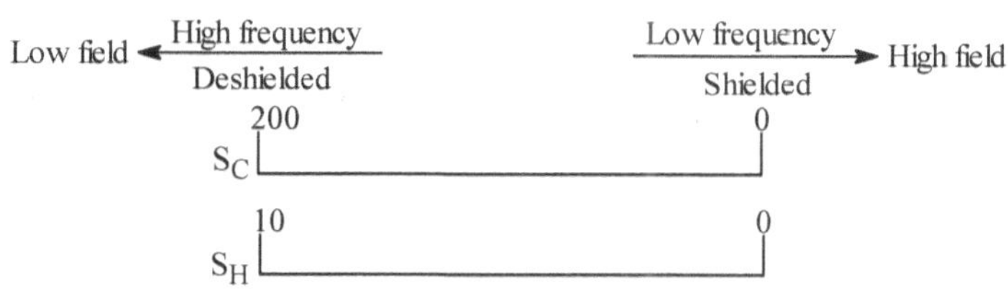

NMR spectrum helps us as:

a) the number of signals, which tells us how many different "kinds" of protons there are in a molecule;

b) the positions of the signals, which tell us something about the environment of each kind of protons;

c) the intensities of the signals, which tell us how many proton of each kind there are, and

d) the splitting of a signal into several peaks, which tells us about the environment of a proton with respect to other nearby protons.

1.3.3 Plasma Emission Spectrometry [46]

The use of plasma as excitation sources for atomic emission is very important in recent years, and inductively coupled plasma (ICP) spectrometer are used for multielement determinations.

The ICP discharge is caused by the effect of a radio frequency field on a flowing gas. Argon gas flows upward through a quartz tube, around which is wrapped a copper coil or solenoid. The coil is energized by a radio frequency generator operating at 5 to 75 MHz and 1 to 2 kW, creating a changing magnetic field in the flowing gas inside the coil. This induces a circulating eddy current in a conductor (the gas), which, in turn, heats it. Argon is not a conductor at room temperature, but it can be made electrically conducting by heating it. To initiate the ICP discharge from Telsa coil or a pilot spark is applied to the flowing argon. The argon is quickly heated, with a stable plasma being produced having a core temperature of about 9000 to 10,000 K.

This type of excitation source has a number of advantages. First, samples can be introduced in solution form through a spray chamber nebulizer. This make quantitative analysis and sample handling much easier . Modern day instruments replace the photographic film detector in an emission spectrograph with a series of photomultiplier tubes set at wavelengths for specific elements (eg., 40 elements), and so quantitative multielement analysis of solutions becomes routine. The plasma is well suited for the refactory(oxide-forming) elements such as boron, phosphorus, uranium, and tungsten and for difficult-to-excite elements such as zinc and cadmium. Detection limits are very competitive with those using atomic absorption.

1.3.4 Atomic absorption spectroscopy [47]

> In this technique the sample solution is aspirated into the flame and the sample element is converted to atomic vapour. The flame then contains atoms of that element. Some are thermally excited by the flame but most remain in the ground state. These ground-state atoms can absorb radiation given off by a special source made from that element. The wavelengths of radiation given off by the source are the same as those absorbed by the atoms in the flame.

The absorption follows Beer's law. That is, the absorbance is directly proportional to the path length in the flame and to the concentration of atomic vapour in the flame. Both of these variables are difficult to determine, but the path length can be held constant and the concentration of atomic vapour is directly proportional to the concentration of the analyte in the solution being aspirated. The procedure used is to prepare a calibration curve of concentration in the solution versus absorbance.

1.4 Literature Review

Metal binding substances, many of which function by chelation, form a class of substances which have furnished many useful drugs and other substances of value in selective toxicity. Sulfonamides have been extensively used in medicine for their antibacterial properties. The combined antibacterial activity of sulfonamides and antimicrobial activity of heavy metals with a view to establish the relationship and importance of metal–drug interactions have been investigated [48,49,50,51and 52]. The metal chelates of sulfa drugs have been found to be more bacteriostatic than the drugs themselves [7and 8]. Sulfamethoxazole (SMX) is used as an agent to treat urinary tract infections. The synthesis and characterization of some first row transition metal complexes of SMX have been reported by Kanagaraj and Rao [53]. In their spectroscopic studies on SMX and its complexes, only the group frequencies have been given in the region 4000–1000 cm^{-1}. They mentioned that SMX acts as a bidendate ligand, binding the metal ion through deprotonated sulfonamide nitrogen and sulfonyl oxygen in the case of Cr, Mn and Ni complexes or isoxazole ring nitrogen and deprotonated sulfonamide nitrogens in the case of Cu complex. SMX probably acts as a monodendate ligand binding the metal ion through deprotanated sulfonamide nitrogen in the case of Fe and Zn complexes [1].

The coordination behavior of sulfamethoxazole complexes of Co and Cd have been reported by Kesimli and Topacli [1] by using IR spectra in the range 4000-400cm^{-1}. They have given the interpretation of all vibrational modes of SMX molecule and also low energy calculation were done by using geometry optimization.They mentioned that SMX coordinates through sulfonamide and amino nitrogens in case of cobalt complex. In case of cadmium complex, the ligand binds the metal ion through sulfonyl oxygen and sulfonamide nitrogen. Fig.1.2 and Fig.1.3 shows the probable structure of Co and Cd complexes in three dimensions, respectively.

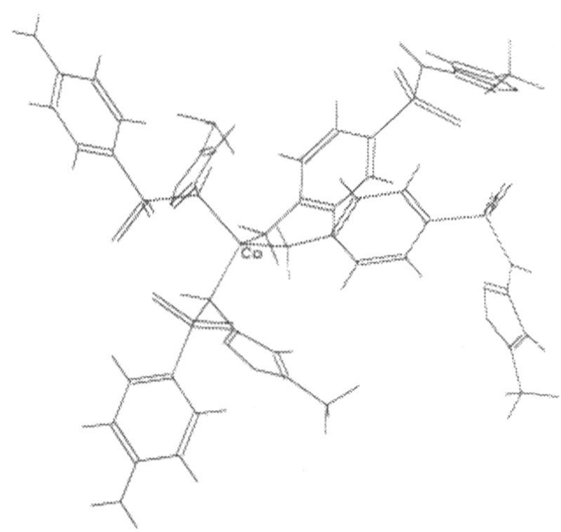

Fig.1.2. The probable structure of Co complex

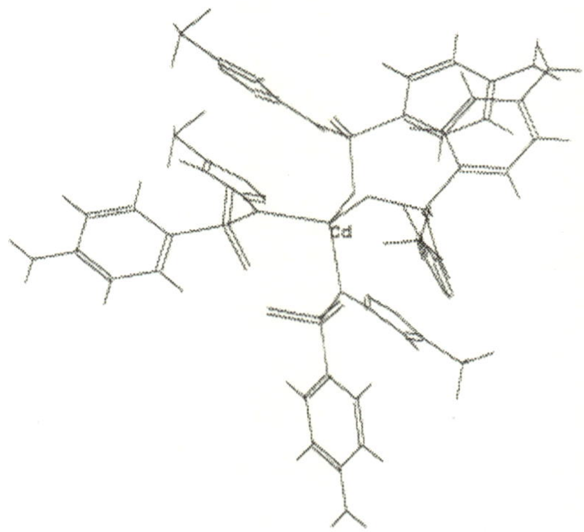

Fig. 3. The probable structure of Cd complex

The SMX wave numbers observed in the IR spectra of the Co and Cd complexes, together with their measurements on free SMX is given in Table 1.3 [1]. The assignments of SMX in the range 4000–1000 cm^{-1} and 1000–400 cm^{-1} taken from Kanagaraj [53] and Jigeesh [54], respectively, were included for comparison.

Table 1.3: The fundamental wave numbers (cm^{-1}) of SMX in the metal complexes

[illegible]	[illegible]	[illegible]	[illegible]	[illegible]
[illegible]	[illegible]	[illegible]	[illegible]	[illegible]
[illegible]	[illegible]	[illegible]	[illegible]	
[illegible]	[illegible]	[illegible]	–	[illegible]
[illegible]	[illegible]	[illegible]	[illegible]	[illegible]
[illegible]	[illegible]	[illegible]	[illegible]	[illegible]
[illegible]	[illegible]	[illegible]	[illegible]	
[illegible]	[illegible]	[illegible]	[illegible]	[illegible]
[illegible]	[illegible]	[illegible]	[illegible]	
[illegible]	[illegible]	–	[illegible]	[illegible]
[illegible]	[illegible]	[illegible]	[illegible]	
[illegible]	[illegible]	[illegible]	–	[illegible]
[illegible]	[illegible]	[illegible]	[illegible]	
	[illegible]	[illegible]	[illegible]	[illegible]
		[illegible]	[illegible]	
–	[illegible]	[illegible]	–	[illegible]
–	[illegible]	[illegible]	[illegible]	[illegible]
–	[illegible]	[illegible]	[illegible]	[illegible]
–	[illegible]	[illegible]	[illegible]	[illegible]
–	[illegible]	[illegible]	[illegible]	[illegible]

The effect of heating on the molecular structure of SMX and its metal complexes using IR spectra in the range between 4000-400 cm^{-} have been reported by Kesimli and Topacli [55] to provide a better understanding of the pharmacological studies. The comparison between the SMX wavenumbers observed in the IR spectra of the heated samples, at different temperatures, and their measurement on crystalline SMX at room temperature helped in defining the temperature values at which the samples are decomposed. A comparison of the main group wavenumbers of SMX observed in the IR spectra of the metal complexes, together with our measurement on crystalline SMX, are given in Table 1.4 [55] at both the room and the decomposition temperatures.

Table 1.4 The fundamental vibrational wavenumbers (cm^{-1}) of SMX in the $M(SMX)_2$ (M=Co and Cd) at room and decomposing temperature

[illegible]		[illegible]		[illegible]		[illegible]
[illegible]	[illegible]	[illegible]	[illegible]	[illegible]	[illegible]	
[illegible]	[illegible]	[illegible]	[illegible]	[illegible]	[illegible]	[illegible]
[illegible]	[illegible]	[illegible]	[illegible]	[illegible]	[illegible]	
[illegible]	[illegible]	[illegible]	[illegible]	-	-	[illegible]
[illegible]	[illegible]	[illegible]	[illegible]	[illegible]	[illegible]	[illegible]
[illegible]	[illegible]	[illegible]	[illegible]	[illegible]	[illegible]	[illegible]
[illegible]	[illegible]	[illegible]	[illegible]	[illegible]	[illegible]	
[illegible]	[illegible]	[illegible]	[illegible]	[illegible]	[illegible]	[illegible]
[illegible]	[illegible]	[illegible]	[illegible]	[illegible]	[illegible]	
[illegible]	[illegible]	-		[illegible]	-	[illegible]
[illegible]		[illegible]	[illegible]	[illegible]	[illegible]	
[illegible]	[illegible]	[illegible]	[illegible]	-		[illegible]
[illegible]		[illegible]	[illegible]	[illegible]	[illegible]	
[illegible]	[illegible]	[illegible]	[illegible]	[illegible]	[illegible]	[illegible]
		[illegible]		[illegible]		
[illegible]	[illegible]	[illegible]	-	-	-	[illegible]
[illegible]	[illegible]	[illegible]	[illegible]	[illegible]	[illegible]	[illegible]
[illegible]	[illegible]	[illegible]	[illegible]	[illegible]	[illegible]	[illegible]
[illegible]	[illegible]	[illegible]	[illegible]	[illegible]	[illegible]	[illegible]
[illegible]	[illegible]	[illegible]	[illegible]	[illegible]	[illegible]	[illegible]

As seen from Table 1.4 the molecular structure of free SMX and Co and Cd complexes are decomposed at 190, 230 and 170°C, respectively. In their previous study [1] of $\nu(NH_2)$, the asymmetric and symmetric vibrations of the aminic group of SMX observed in the IR spectrum of the Co complex had shifted upon coordination towards a lower wavenumber compared with the free SMX. As seen from Table 1.4, the observed wavenumbers of $\nu_{as}(NH_2)$ and $\nu_s(NH_2)$ in the IR spectrum of the heated Co complex approach their position in the IR spectrum of free SMX.

According to this result, it is considered that the Co complex tends to break from the NH_2 group due to the effect of higher temperature. In their previous study [1], the bands observed at 1157 (1144 cm^{-1}) and 1317 cm^{-1} (1308 cm^{-1}) had been assigned to the symmetric and asymmetric stretching modes of the SO_2 group in the spectrum of free SMX, respectively. In the IR spectrum of the Cd complex, the symmetric stretching band of SO_2 observed at 1157 cm^{-1} had disappeared and another band observed at 1144 cm^{-1} showed a shift towards a lower wave number (about at 1110 cm^{-1}). This conclusion was based on the fact that the coordination in the Cd complex occurs through the sulfonyl oxygen. In the IR spectrum of heated SMX, $\nu_{as}(SO_2)$ and $\nu_s(SO_2)$ modes have upward shifts in wave number compared to those in the free molecule at room temperature. Also, the splitting of these bands disappears. As seen from Table 1.4, the observed wave number of $\nu_s(SO_2)$ band at 1124 cm^{-1} in the IR spectrum of the heated Cd complex approaches its position in the IR spectrum at room temperature. So it is considered that the molecular structure of the Cd complex tends to break from the SO_2 group. In addition, in their previous study [1], the $\nu(SN)$ vibration observed at 929 cm^{-1} in the IR spectrum of SMX was found to have splitted and shifted to higher wavenumbers in the spectra of metal complexes. This result had suggested that SMX also bonded with the metal ions through the sulfonamide nitrogens. As seen from Table 1.4, this band appeared as a singlet band and showed a shift towards a lower wavenumber. So the ν(SN) vibration approaches its position and shape in the IR spectra of free SMX at room temperature. According to these results, it is considered that the Co and Cd complexes also tend to break from the SN group due to the effect of higher temperature. The vibrational analysis and calculations indicated that the stabilities of metal complexes are in the order Co>Cd which are in agreement with the increasing order of the second ionization potential of these metals.

1.5 The Aim of the Study

So far, a few experimental studies have been reported for sulfanilamide and its metal complexes. Narang and Gupta [53] and Blasco et al. [54] had investigated some IR group frequencies of sulfanilamide to determine its possible transition metal binding sites. The coordination behavior of sulfamethoxazole complexes of Co and Cd have been reported by Kesimli and Topacli [1] by using IR spectra in the range 4000-400cm^{-1}.

Sulfonamides (sulfamethoxazole (SMX), sulfadimethoxine (SDMX), sulfadiazine etc.) are well-known antibacterial drugs. Heavy metals are essential for all forms of life. The antimicrobial activity of heavy metals make them more important in the field medicine. The combination of antimicrobial and antibacterial activities of sulfonamides and heavy metals increased their importance in the field of metal complexes of sulfonamides [55]. the metal complexes of sulfa drugs have been found to be more bacteriostatic than the drugs themselves [7 and 8].

As an extention of previous studies on sulfamethoxazole and its complexes, we synthesized new sulfamethoxazole complexes of palladium, iridium and tin and investigated the coordination behaviour of the drug by using FT-IR spectra in the range 4000-400 cm^{-1}, UV-visible spectroscopy, AAS, ICP-AES and NMR. Antibacterial activity of these metal complexes have also been checked and reported in this work. It is observed that antibacterial activity of sulfamethoxazole is increased on chelation.

We also synthesized new pyrimethamine complexes of zinc and cadmium and same above technique were used for investigating the coordination behaviour of the drug.

EXPERIMENTAL

2.1 Glassware and Chemicals

Glassware used during the present investigation was of Pyrex. Solvents such as tetrahydrofuran (THF), dichloromethane, diethylether toluene, ethanol, acetone and methanol were procured from E-Merck (Germany). These all were dried before use by standard analytical methodology [56]. $FeCl_2.4H_2O$, $PClCl_2$, $IrCl_3.3H_2O$, $SnCl_2$, sulfamethoxazole and pyrimethamine were also of E-Merck origin.

2.2 Instrumentation

Electrochemical melting point apparatus, Model Mp.D Mitamura Riken Kogyo Japan was used for determining melting points. Infrared spectra were recorded on KBr disc on Perkin Elmer FT-IR Model (Spectrum 1000) Spectrometer. NMR spectra were recorded on Bruker NMR-300 MHz. For metal estimation UV-Visible Spectrophotometer 1700 Shimadzu was used.

Inductively Coupled Argon Plasma Atomic Emission Spectrometer Thermo Jarrell Ash, IRIS was used for the determination of metallic content in the organomeallic compounds.

2.3 Preparation of Metal Sulfamethoxazole Complexes

In literature Co and Cd complexes of sulfamehoxazole (SMX) were prepared in the following manner: Sulfamethoxazole (2 mmol) was dissolved in hot water. Then dilute sodium hydroxide solution was added with constant stirring. The ligand solution (pH=8–10) was mixed with an aqueous solution of the metal acetate (1 mmol) under continuous stirring. The compounds, which precipitated, were filtered, washed with distilled water and dried at room temperature for a few days [1]. In this study aqueous solution of metal chlorides have been used and Sulfamthoxazole have been dissolved in organic solvent like acetone, methanol, THF and ehanol.

2.3.1 Synthesis of Palladium Sulfamethoxazole Complex

Palladium(II) chloride (0.1g, 0.564 mmole) dissolved in 25 mL water. 4-Amino-N-(5-methyl-3-isoxazolyl)benzenesulfonamide (0.111g, 0.56 mmole) solution was prepared in acetone. Sulfamethoxazole solution was added dropwise by dropping funnel with constant stirring into $PdCl_2$ solution. Light yellow precipitates were obtained immediately. Reaction mixture was kept on stirring for one hour. Precipitates were filtered and washed with ether and dried in a vacuum desiccator.

2.3.2 Synthesis of Iron Sulfamethoxazole Complex

Iron(II) chloride tetrahydrate (0.1g, 0.502 mmole) was dissolved in 25 mL ethanol on heating sulfamethoxazole (0.127g, 0.564 mmole) solution was prepared in ethanol and was added dropwise by dropping funnel along with constant stirring into $FeCl_2.4H_2O$ solution and was refluxed for 7 hours. The reaction mixture was stirred for 3 days and turned turbid. Solvent was evaporated on rotary evaporator. Dark brown solid was obtained.

2.3.3 Synthesis of Stannous Sulfamethoxazole Complex

Tin(II) chloride (0.1g, 0.527 mmole) was dissolved in 25 mL THF. Sulfamethoxazole (0.13g, 0.527 mmole) solution was prepared in THF and was added dropwise in $SnCl_2$ solution refluxed for one hour and stirred for 2 days, solution turned light yellow. Solvent was evaporated on rotary evaporator which resulted into powdered form by addition of dichloromethane.

2.3.4 Synthesis of Iridium

Iridium(III) chloride trihydrate (0.1g, 0.283 mmole) was dissolved in 25 mL of ethanol on heating. Sulfamethoxazole (0.071g, 0.283 mmole) solution was prepared in ethanol on heating. Maroonish brown solution turned to dark purple after 1 hour reflux and 24 hour stirring. Dark purple precipitates were obtained which were filtered and dried under vacuum.

2.4 Preparation of Metal Pyrimethamine Complexes

2.4.1 Synthesis of Cobalt Pyrimethamine Complex

Cobalt(II) chloride hexahydrate (0.1g, 0.42 mmole) was dissolved in 30 mL of THF. Pyrimethamine (0.098g, 0.42 mmole) solution was prepared in THF on heating. Dropwise addition of pyrimethamine solution into blue coloured. $CoCl_2.6H_2O$ solution resulted to dirty green. Precipitates were filtered and dried under vacuum.

2.4.2 Synthesis of Cadmium Pyrimethamine Complex

Cadmium(II) chloride dehydrate (0.1g, 0.45 mmole) was dissolved in 30 mL ethanol. Pyrimethamine (0.2g, 0.900 mmole) solution was prepared in ethanol in heating and was added dropwise in $CdCl_2.2H_2O$ solution which turned turbid immediately. Reaction mixture was stirred for 24 hours. The solvent was evaporated on rotary evaporator and white powder was collected.

2.4.3 Synthesis of Zinc Pyrimethamine Complex

Zinc(II) chloride dehydrate (0.1g, 0.58 mmole) was dissolved in 30 mL ethanol. Pyrimethamine (0.111g, 0.58 mmole) solution was prepared in ethanol on heating and was added dropwise in $ZnCl_2.2H_2O$ solution which turned turbid. The solvent was evaporated on rotary evaporator and off white powder was collected.

2.5 Characterization of Metal Complexes

2.5.1 Physical Properties

Metal pyrimethamine complexes and metal sulphamethoxazole complexes, like many other coordination compounds, containing organic ligands, not have sharp melting points. Solubility of these metal complexes were also tested in different solvents. A complete data on physical properties of metal pyrimethamine complexes and sulfamethoxazole is given in Table 2.1.

2.5.2 Estimation of Metals in Metal Sulfamethoxazole Complexes

The complex iron sulfamethoxazole was analyzed for its metal ion content by spectroscopic method and Inductively Coupled Plasma Atomic Emission Spectrometer (ICP-AES). While palladium was estimated gravimetrically along with above two techniques mention. These complexes were first decomposed with nitric acid and their solutions were used for the estimation of metal ion.

An accurately weighed sample of the complex was treated with 10 ml of concentrated nitric acid. The mixture was heated on a Bunsen burner and evaporated to a paste. To the paste additional 10 ml of nitric acid was added. The reaction mixture was heated for complete decomposition. The residue was dissolved in distilled water and then diluted to a known volume in a volumetric flask. Calibration curves were drawn using standard solutions of respective metal salts. Absorption of metal solutions were also measured and their concentrations were obtained from calibration curve as stated below:

Estimation of Palladium [57]

Palladium in the complex was estimated by a reported spectrophotometric method using dimethylglyoxime.

A stock solution (100 ppm) of Pd^{+2} was prepared by dissolving 8.349×10^{-3}g of $PdCl_2$ in distilled water in 50 mL flask. 2, 4, 6 and 10 ppm solutions were obtained by diluting 1, 2, 3, 4, 5 mL portions of 100 ppm solution, respectively. 10 mL of aliquots of these solutions were taken in 25 mL volumetric flasks. To each of these solutions, 2 mL of dimethylglyoxime solution was added and finally diluted to 25 mL with distilled water. These standard solutions were used too for making a calibration curve by plotting absorbance versus metal ion concentration. The palladium sulfamethoxazole complex was decomposed with nitric acid, colour was developed in the same manner as for the above standard palladium solution. The absorbance of both sample and standards were measured on Shimadzu spectrophotometer at λ_{max} 476 nm. The concentration of palladium was determined in test solution using the calibration curve.

Observation

Absorbance values of standard solutions of Pd(II) at $\lambda_{max} = 476$ nm.

Table 2.1

S.No.	Metal Concentration	Absorbance
1.	0	0.00
2.	4	0.128
3.	6	0.193
4	8	0.269
5.	10	0.331
6.	Unknown conc. of Pd soln.	0.294

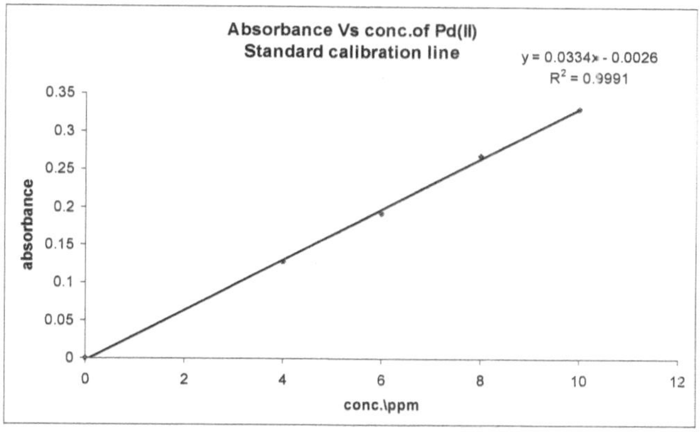

The unknown concentration from calibration curve is 8.9 ppm.

Calculation

Unknown concentration from curve
= 8.9 ppm/25 mL

$M_1V_1 = M_2V_2$

$M_1 \times 10 = 8.9 \times 25$

$M_1 = 8.9 \times 25/10 = 22.25$ ppm/50 mL

Amount of Palladium Sulfamethoxazole

Complex taken
= 0.005g

Amount of palladium found
= 0.0011g

%age of palladium
=22%

Gravimetrically Estimation of Palladium Using Dimethylglyoxime [58]

An accurately weighed amount (0.0400g) of palladium sulfamethoxazole complex was decomposed with concentrated HNO_3 to near dryness. The resulting paste was dissolved in distilled water, volume was made to 100 mL in a volumetric flask.

1% solution of dimethylglyoxime in 95% ethanol was added to it and the solution was allowed to stand for one hour. Orange yellow coloured precipitates of palladium dimethylglyoxime complex precipitated out. It was filtered through a weighed sintered glass crucible No.ASTM 36060. The precipitate was washed with cold water first, then with hot water and was dried to constant weight at 110°C. From the weight of palladium complex $[Pd(C_4H_7O_2N_2)_2]$, percentage of palladium in palladium penicillin G complex was calculated.

Amount of the palladium complex taken
= 0.0400g

Amount of the palladium found
= 0.0084g

Percentage of palladium
= 21%

Estimation of Iron with 2,2′-Bipyridyl

Iron in the complex was estimated by a reported spectrophotometric method using 2,2′-bipyridine.

A stock solution (1000 ppm) of Fe^{+2} was prepared by dissolving 0.35 of $Fe(SO_4).(NH_4)_2SO_4.6H_2O$ in distilled water. In 50 mL flask. 20, 40, 60, 80 and 100 ppm solutions were obtained by diluting 1, 2, 3, 4 and 5 mL portions of 1000 ppm solution respectively. 2 mL of aliquots of these solutions were taken in 25 mL volumetric flask. To each of these solutions 1 mL of 1% 2,2′-bipyridyl was added and diluted upto the mark, with distill water.

These standard solutions were used for making a calibration curve by plotting absorbance versus metal ion concentration.

The iron sulfamethoxazole complex was decomposed with nitrice acid. Colour was developed in the same manner as for the above standard iron solution.

The absorbance of both sample and standards were measured on spectrophotometer at λ_{max} 522 nm. The concentration of iron was determined in test solution using the calibration curve.

Observation

Absorbance values of standard solutions of Fe(II) at λ_{max} = 522 nm.

Table 2.2

S.No.	Metal Concentration	Absorbance
1.	0	0.00
2.	20	0.085
3.	40	0.179
4	60	0.266
5.	80	0.371
6.	100	0.455
7.	Unknown	0.024

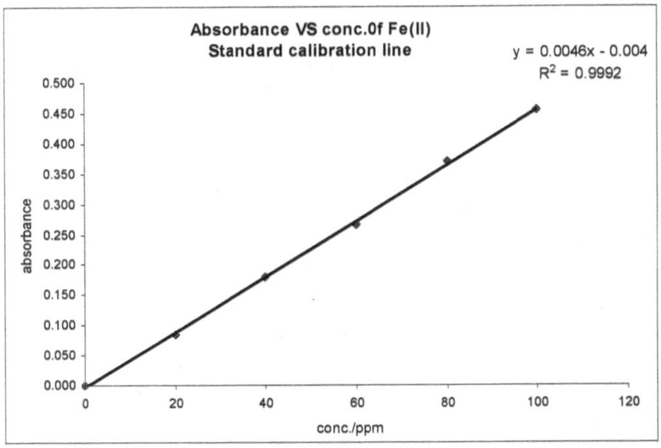

The unknown concentration from calibration curve is 6.2 ppm.

Calculation

Unknown concentration from curve
= 6.2 ppm/25 mL

$M_1V_1 = M_2V_2$

$M_1 \times 2 = 6.2 \times 25$

$M_1 = 6.2 \times 25/2 = 77.5$ ppm/50 mL

$77.5 \times 50/1000 = 3.87$ mg of Fe^{+2}

Amount of Iron complex taken
= 0.03g

Amount of Fe^{+2} found

$$= \frac{3.87 \times 10^{-3}}{0.03} \times 100$$

= 12.9%

Estimation of Zinc

Zinc in the complex was estimated by using atomic absorption spectroscopy technique.

For standard solutions were prepared of $ZnCl_2.2H_2O$ for obtaining standard calibration line at λ_{max} 213.9 nm. Zinc complex (0.00599g) was digested with HNO_3 and solution was made in 10 mL volumetric flask, and absorbance was noted.

Amount of Zinc complex taken
= 0.00599g

Amount of zinc found
= 7.4812×10^{-4}g

%age of zinc

= 12.5%

2.5.3 Estimation of Metals Using Inductively Coupled Argon Plasma Atomic Emission Spectrometer

Organometallic samples were digested with the help of concentrated nitric acid and perchloric acid. An accurately weighed sample (0.1g) was placed into a 50 ml beaker and treated with 10 ml of concentrated nitric acid. The contents of the beaker were then heated and treated with 3 ml of perchloric acid. The digestion was continued until dense white fumes evolved marking the completion of the dissolution process. The contents of the flask were transferred to a 50 ml volumetric flask and the volume was made up to the mark with doubled distilled water. A blank, without sample, was also prepared in the same way. Linear calibration method of quantification was used to quantify the results. An inductively coupled argon plasma atomic emission spectrometer (Thermo Jarrell Ash, IRIS) was used for the determination of metallic content in the organometallic compounds by the following analytical conditions.

RF power
= 1150 W

Flash pump rate
= 100 rpm

Analysis pump rate
= 100 rpm

CID integration time (low WL)
= 15 sec

CID integration time (high WL)
= 5 sec

% of Pd determined

= 23.9%

% of Fe determined

= 14.5%

2.6 Antibacterial Assay

The agar well diffusion method was used to screen for antibiotic activity.

2.6.1 Preparation of Samples

A total 1 mg of the complexes was dissolved in 1 mL of DMSO to get 1 mg/mL concentration. Solution of Roxithromycin and Cefixime-USP, 1mg/mL in DMSO, were prepared for positive control. Pure DMSO was used as negative control.

2.6.2 Media for Bacteria

1. Nutrient broth medium (Merck) was used to grow bacteria for incculums preparation. Its composition was:

   a) Peptone from meat
   = 5 gm/L

   b) Meat extract
   = 3 gm/L

2. Nutrient agar medium (Merck) was used to perform antibacterial assay. Its composition was:

   a) Peptone from meat
   = 5 gm/L

   b) Meat extract
   = 3 gm/L

   c) Agar-agar
   = 12 gm/L

2.6.3 Preparation of Media for Bacteria

Nutrient broth medium was prepared by dissolving 8.0 gm/L nutrient broth in distilled water; pH was adjusted at 7.0 and was autoclaved. Nutrient agar medium was prepared by dissolving 20 gm/L of nutrient agar in distilled water; pH was adjusted to 7.0 and was autoclaved.

2.6.4 McFarland 0.5 $BaSO_4$ Turbidity Standard

The standard was prepared by adding 0.5 mL, 0.048M $BaCl_2$ to 99.5 mL, 0.36N H_2SO_4. Barium sulfate turbidity standard (4 to 6 mL) was taken in screw capped test tube and was used to compare the turbidity (Koneman, 1988).

2.6.5 Microorganisms Used

Six strains of bacteria were used, one was gram positive, which was *Micrococcus luteus* (ATCC 10240) and five were gram negative, which were *Escherichia coli* (ATCC 15224), *Salmonella setubal* (ATCC 19196), *Pseudomonas picketii* (ATCC 49129), *Bordetella bronchiseptica* (ATCC 4617) and *Enterobacter aerogens* (ATCC 13048). The organisms were maintained on nutrient agar medium at 4°C.

2.6.6 Preparation of Inocula

Centrifuged pellets of bacteria from twenty-four hours old culture in nutrient broth (Sigma) of selected bacterial strain was mixed with physiological saline and turbidity was corrected by adding sterile physiological saline until a McFarland 0.5 $BaSO_4$ turbidity standard [10^8

colony forming unit (CFU) per mL]. This inoculum was used for seeding the nutrient agar.

2.6.7 Preparation of Seeded Agar Plates

Nutrient agar medium was prepared by suspending nutrient agar (Merck) 20 gm in 1 liter distilled water; pH was adjusted at 7.0 and was autoclaved. It was allowed to cool up to 45°C. Then it was seeded with 10 mL of prepared inocula to have 10^6 CFU per mL. Petri plates (14 cm) were prepared by pouring 75 mL of seeded nutrient agar and allowed to solidify. Eleven wells per plate were made with sterile cork borer (8 mm).

2.6.8 Pouring of Test Solutions, Incubation and Measurement of Zone of Inhibition

Using Micropipette, 100 μL of test solutions was poured in respective well. Eight concentrations of extract, two solutions for positive control (Roxithromycin and Cefixime-USP), one for each) and one for negative control (DMSO) was applied to each Petri plate. These plates were incubated at 37°C. After 24 hours and 48 hours of incubation, the diameter of the clear zones showing no bacterial growth around each well was measured. Triplicate plates were prepared for each strain. Mean clear zone of these plates was calculated with standard deviation. Antibacterial activity of all dilutions of extract was determined against six bacterial strains.

The results are reproduced in Table 3.10 and 3.11.

RESULTS AND DISCUSSIONS

Present study involves synthesis of some new metal complexes with their characterization using spectroscopic techniques and studying their biological application.

A number of metallic complexes of sulfamethoxazole with Pd(II), Fe(II), Sn(II) and Ir(II) were synthesized from methanol-aqueous medium in 1:1 ration.

General scheme of reaction as follow.

$MCl_2 + L \longrightarrow MLCl_2$

M = Pd(II), Ir(III), Sn(II), Fe(II),
L = Sulfamethoxazole

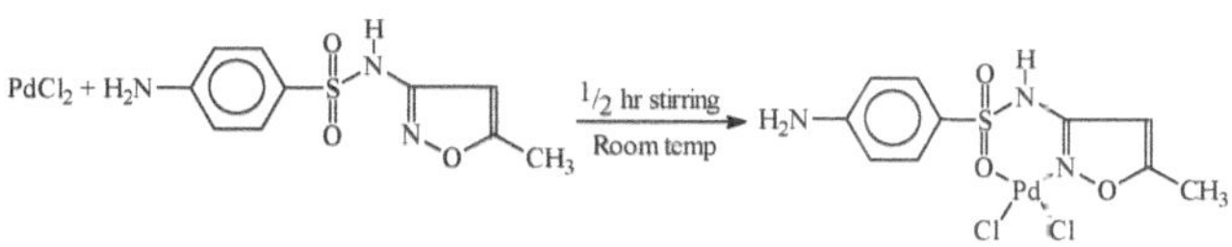

Similalry number of metallic complexes of pyrimethamine with Zn(II), Co(II) and Cd(II) were synthesized from methanol-aqueous. Medium in 1:1 molar ratio.

$$ZnCl_2.2H_2O + C_{12}H_{13}ClN_3 \xrightarrow[\text{1/2 stirring}]{\text{1/2 hr reflux}} Zn(C_{12}H_{13}ClN_4)Cl_2.4H_2O$$

These complexes were characterized by elemental analysis and instrumental techniques such as FT-IR, UV-visible spectrophotometer, atomic absorption spectroscopy, ICP-ES inductively coupled plasma emission spectroscopy, ^{1}H-NMR, ^{13}C-NMR and mass spectrometry.

Antibacterial activity of these complexes against a number of gram-positive and gram-negative bacteria have been checked and compared with the ligands.

Physical data of these complexes is given in Table 3.1 and 3.2. Both the ligands sulfamethoxazole and pyrimethamine are white crystalline whereas metal complexes of these ligands are coloured. Generally, all complexes are amorphous. The decomposition points of these complexes vary in the range 90-310°C. The complexes are insoluble in water but easily dissolve in dimethylsulfoxide, their various solubility in various organic solvents are given in Table 3.1 and 3.2.

Table 3.1: Physical data of SMO and synthesized metal complexes.

SN	Complex	Color	Physical Form	MP (°C)	Mol.Wt	Solubility
1	SMOZ	White	crystalline	275	253.28	THF(s), Ethanol(s) Acetone(s)
2	Palladium sulfamethoxazole	Orange	Amorphous	290-310	430	DMSO(s)
3	Iron sulfamethoxazole	Dark brown	Amorphous	120-130	379.81	DMSO(s), THF(ps), Acetone(ps), Ethanol (ps)
4	Iridium sulfamethoxazole	Purple	Amorphous	138-150	551.6	DMSO(s), THF(s), Acetone(s), Ethanol (s)
5	Tin sulfamethoxazole	Light yellow	Amorphous	90-110	442.59	DMSO(s), THF(s), Acetone(s), Ethanol (s)

Table 3.2: Physical data of PMA and synthesized metal complexes.

Sr.No	Complex	Colour	Physical form	MP (°C)	Mol.Wt	Solubility
1	PMA	White	crystalline	237	248.7	THF(s), Ethanol(s) Acetone(s)
2	Zinc pyrimethamine	Off white	Amorphous	210-212	455.88	DMSO(s), Ethanol(s), $CHCl_3$(ps)
3	Cobalt pyrimethamine	Bluish green	Amorphous	206-208	627.3	DMSO(s), $CHCl_3$(ps), Ethanol(s)
4	Cadmium pyrimethamine	White	Amorphous	280-286	680.8	DMSO(s), THF(s), Ethanol(s)

3.1 FT-IR Spectra

3.1.1 Metal Sulfamethoxazole Complexes

The ligand, sulfamethoxazole has a number of electron pair donor atoms, such as N, O, S which can bind with transition metal ions through different possible coordinate sites. Infrared spectra can be used for finding the coordination sites of these metal complexes. In case of sulfamethoxazole the representative IR bands of the group which are attached to metal, are expected to either shift or completely disappear [61].

The infrared spectra of solid ligand and its complexes have been recorded in KBr phase. The major band positions along with their assignments are reported in Table 3.3. The infrared spectra of the complexes may be compared with the ligand and tentative assignments about metal ligand binding side may be made. The assignments of SMX in the range 4000–1000 cm^{-1} and 1000–400 cm^{-1} taken from Kanagaraj [53] and Jigeesh [54], respectively, were included for comparison.

Table 3.3: Principle IR bands (cm^{-1}) of sulfamethoxazole and its complexes.

Sr.No	SMOZ	Pd^{+2} complex	Fe^{+2} complex	Ir^{+3} complex	Sn^{+2} complex	Functional groups
1	3466 3377 3298 3142	3553 3489 3223	3474 3385 3109	3476 3387 3235 3158	3475	N–H C–H Stretching
2	2931	–	2925 2856	2926 2857	–	$-CH_3$
3	1621	1612	1593	1610	1615	C=N
4	1595 1471	1489	1593 1464	1499	1497	Aromatic C=C
5	1366	–	1387	1383	1389	S=O(asym)
6	1144	1144	1159	1161	1164	S=O(sym)
7	685	678	685	679	679	C–S

The study of IR spectra of the complexes show that C=N of sulfamethoxazole at 1621 cm^{-1} is lowered in all complexes. Kanagaraj and Rao [53] had observed these band at 1620. This frequency is shifted to 1612, 1593, 1610 and 1615 cm^{-1} in palladium(II), iron(II), iridium(III), tin(II) respectively. This is indicative of the fact that C=N is involved in bonding directly or indirectly.

Kanagaraj and Rao [53] had assigned the bands at 1320 (1310) cm^{-1} and 1155 (1145) cm^{-1} to antisymmetric and symmetric modes of SO_2 group of SMX. In our study ν_{asym} S=O of the ligand 1366 cm^{-1} was not observed in palladium(II) complexes while it was shifted to 1387, 1383 and 1389 cm^{-1} in iron(II), iridium(III) and tin(II) respectively indicating that oxygen of sulfonamide group has direct participation in bonding with metal.

At 685 cm^{-1} there appears a band attributed to ν_{C-S} stretching vibrations. Jigeesh [54] had assigned the band at 684.This band is not affected upon complexation. The band appears in the same range in all these complexes. The ν_{C-S} stretching vibrations in complexes of palladium(II), iron(II), iridium(III) and tin(II) are observed at 678, 685, 679 and 679 cm^{-1} respectively. This indicates that sulphur is not involved in bonding with these metal ions. The broad band appearing in the range 3466-3142 cm^{-1} in all complexes is attributed to stretching vibrations of C–H and N–H groups. Kanagaraj and Rao [53] had observed these bands at 3467 -3292 cm^{-1}.

From these observations, it is concluded that nitrogen of the ring and oxygen of sulfonamide group are acting as chelating atom forming 6 membered ring as shown.

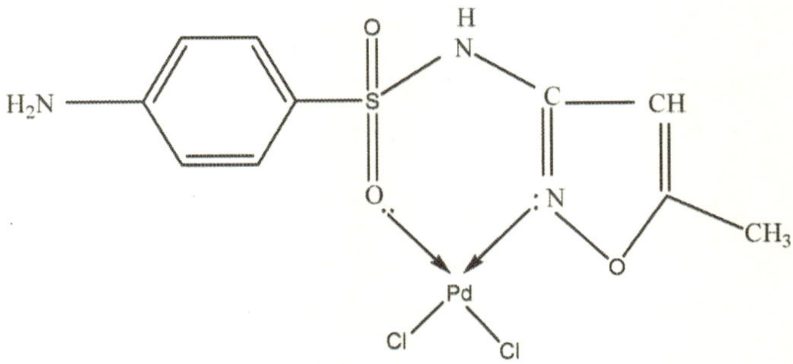

Fig. 3.1: Proposed structure for palladium sulfamethoxazole.

3.1.2 Metal Pyrimethamine Complexes

The ligand pyrimethamine has 4 nitrogen atoms which can donate electron pair two nitrogen of diamine and two conjugated in ring. The major band positions along with their assignments are reported in Table 3.4.

Table 3.4: Principle IR bands (cm^{-1}) of pyrimethamine and its complexes.

Sr.No	PMA	Co^{+2} complex	Zn^{+2} complex	Cd^{+2} complex	Functional group
1	3468 3310 3144	3553 3304 3136	3480 3391 3308 3148	3479 3395 3309 3162	N–H C–H Stretching
2	2975 2936 2878	2965	2975	2977	$-CH_3$
3	1570	1531	1558	1556	C=N Conjugated cyclic system
4	1474	1492	1480	1479	C–N

The study of IR spectra of complexes shows that N–H stretching frequency of pyrimethamine at 3468 cm^{-1} is increased in all the complexes. This N–H frequency is shifted to 3553, 3480, 3479 cm^{-1} in cobalt(II), zinc(II) and cadmium(II) respectively. This is indicative of the fact that $-NH_2$ is involved in banding with metal ions.

$\nu_{C=N}$ of conjugated cyclic system of the ligand at 1570 cm^{-1} is lowered in all the complexes, shifted to 1531, 1558, 1556 cm^{-1} in cobalt(II), zinc(II), cadmium(II) respectively. This is an indication of bonding of metal ion through nitrogen of conjugated cyclic system.

ν_{C-N} of the ligand 1479 cm^{-1} was shifted to 1492, 1480, 1479 cm^{-1} in complexes of cobalt(II), zinc(II) and cadmium(II) respectively, indicating that $-NH_2$ group has direct participation in bonding with metal ions.

From these observations, it is concluded that nitrogen of secondary amine and nitrogen of conjugated ring act as binding sites with metal ion as shown.

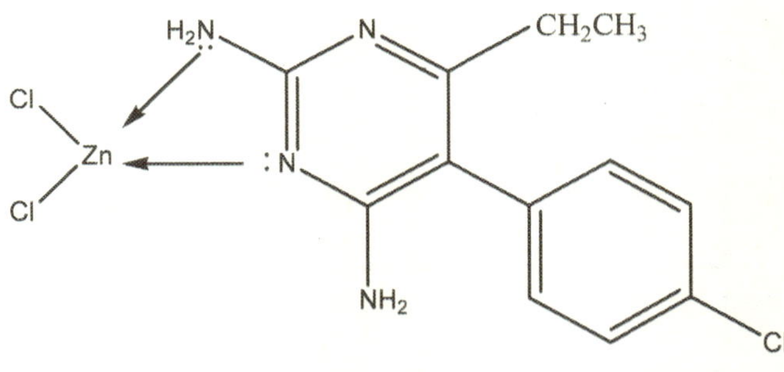

Fig. 3.2: Proposed structure for zinc pyrimethamine.

3.2 ^{1}H NMR Spectroscopy

Nuclear magnetic resonance (NMR) is a spectroscopic method which is even more important to the chemists than infrared spectroscopy. Many nuclei may be studied by NMR techniques, but hydrogen and carbon are most commonly available. NMR gives information about the number of magnetically distinct atoms of the type being studied. When hydrogen nuclei (protons) are studied, for instance, one can determine the number of each of the distinct types of hydrogen nuclei as well as obtain information regarding the nature of their immediate environment [60].

3.2.1 Sulfamethoxazole and Palladium Sulfamethoxazole Complex

In ^{1}H NMR spectrum of palladium sulfamethoxazole complex C–H($H_{(2)}$) resonance has shifted downfield in chelated sulfamethoxazole ligand by 3.43 ppm as compared to free ligand which indicates that due to chelation through nitrogen electron density is decreased on nitrogen thus $H_{(2)}$ is deshielded.

The –NH peaks are often very broad and weak without any distinct coupling to hydrogens on an adjacement carbon atom [61]. $H_{(6)}$ has shifted downfield by 6.1 ppm while $H_{(4)}$ also shifted downfield by 7.47-7.45(d) ppm as compared to free ligand.

Shielding effect observed in $H_{(3)}$ by 6.59-6.56(d) ppm is due to resonance that removes electron density from nitrogen and donates to the ring. This indicates that lone pair of electron of $-\ddot{N}H_2$ are not being utilized for bonding. ^{1}H NMR results support the binding site as proposed by FT-IR results.

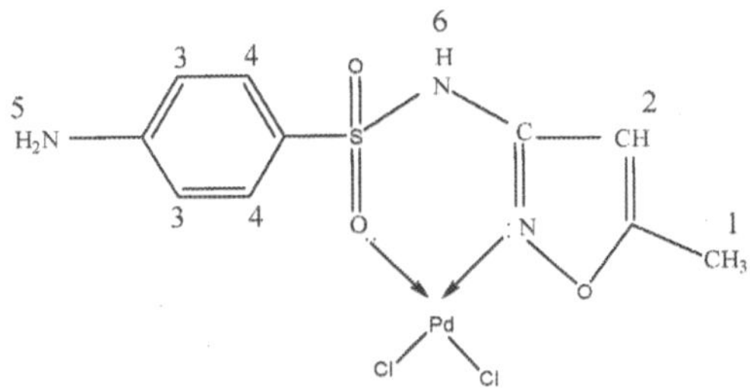

Fig. 3.3: Proposed structure for ^{1}H NMR.

Table 3.5: Comparison of ^{1}H NMR spectra of SMOZ and its Pd^{+2} complex.

Sr.No	Protons	Free ligand SMOZ δ(ppm)	Pd+2 complex δ(ppm)
1	H(1)	2.22(s)	2.28(s)
2	H(2)	3.31(s)	3.43(s)
3	H(3)	6.66–6.63(d)	6.59–6.56(d)
4	H(4)	7.38–7.41(d)	7.47–7.45(d)
5	H(5)	10.93	10.97
6	H(6)	6.0	6.1

3.3 ^{13}C NMR Spectroscopy

The study of carbon nuclei through nuclear magnetic resonance (NMR) spectroscopy is an important technique for determining the structures of organic molecules. Carbon spectra can be used to determine the number of nonequivalent carbons and to identify the types of carbon atoms. Thus ^{13}C NMR provides direct information about the carbon skeleton of a molecule [60].

3.3.1 Sulfamethoxazole and Palladium Sulfamethoxazole Complex

The chemical shifts of ^{13}C NMR are given in the Table 3.2. The carbon atoms C_3, C_4, C_5 and C_6 of free ligand give signals at 158.4, 170.4 and 124.5 ppm. After complexation with palladium(II) it appears at 165.7, 172.7 and 128.0 ppm respectively.

On coordination with Pd^{+2} a downfield shift was observed for C_5 and C_4. The downfield shift is attributed to the shift of electron density from ligand to metal ion. Whereas slight shielding effect was observed for C_1 which attribute to the shift of electron density from $-\ddot{N}H_2$ to C_1 of the ring.

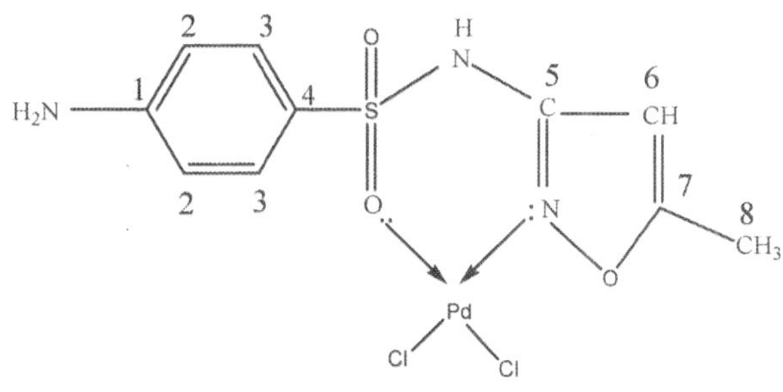

Fig. 3.4: Proposed structure for ^{13}C NMR.

Table 3.6: Comparison of ^{13}C NMR spectra of SMOZ and its Pd^{+2} complex.

Sr.No	Carbon-13	Free ligand SMOZ δ(ppm)	Pd^{+2} complex δ(ppm)
1	C1	129.3	128.9
2	C2	153.8	152.6
3	C3	158.4	165.7
4	C4	170.4	172.7
5	C5	124.5	128.0
6	C6	113.0	114.8
7	C7	112.8	99.0
8	C8	12.5	12.7

Conclusion drawn from NMR spectra study of the complexes and free ligand sulfamethoxazole shows that shift observed in C_3, C_4 and C_5 confirm the formation of complex and support the proposed structure.

3.4 ^{1}H NMR of Pyrimethamine and Zinc Pyrimethamine Complex

In ^{1}H NMR spectrum of zinc pyrimethamine complex N–H($H_{(5)}$) has shifted downfield in chelated pyrimethamine ligand by 3.62 ppm as comparec to free ligand which indicates that electron density from nitrogen is donated to metal ion decrease in electron density on nitrogen as a result deshielding the $H_{(5)}$ attached to nitrogen.

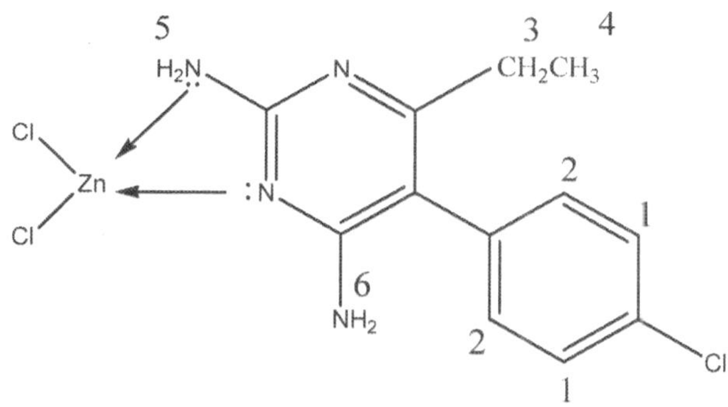

Fig. 3.5: Proposed structure for ^{1}H NMR.

Table 3.7: Comparison of ^{1}H NMR spectra of PMA and its Zn^{+2} complex.

Sr.No	Protons	Free ligand PMA δ(ppm)	Zn+2 complex δ(ppm)
1	H(1)	7.21-7.17(d)	7.21-7.18(d)
2	H(2)	7.49-7.45(d)	7.48-7.45(d)
3	H(3)	2.13-2.05(q)	2.10-2.05(q)
4	H(4)	0.95(t)	0.95(t)
5	H(5)	3.36	3.62
6	H(6)	5.63	5.70

3.5 ^{13}C NMR of Pyrimethamine and Zinc Pyrimethamine Complex

The chemical shifts of ^{13}C NMR are given in the Table 3.8. The carbon atoms C_1 and C_8 of free ligand give signals at 162.6 and 162.5 ppm. After complexation with zinc(II) it appears at 164.0 and 163.4 ppm respectively.

On coordination with Zn^{+2} a downfield shift was observed for C_7 and C_8. The downfield shift is attributed to the shift of electron density from ligand to metal ion.

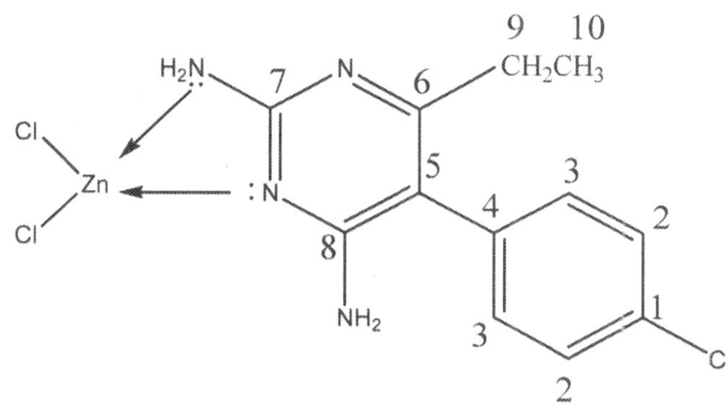

Fig. 3.6: Proposed structure for ^{13}C NMR.

Table 3.8: Comparison of ^{13}C NMR spectra of PMA and its Zn^{+2} complex.

Sr.No	Carbon-13	Free ligand PAM δ(ppm)	Zn^{+2} complex δ(ppm)
1	C1	133.0	133.1
2	C2	132.3	132.4
3	C3	129.4	129.5
4	C4	135.5	135.7
5	C5	105.7	105.5
6	C6	166.9	166.8
7	C7	162.6	164.0
8	C8	162.5	163.4
9	C9	27.9	27.9
10	C10	13.5	13.5

Conclusion drawn from the NMR spectra of pyrimethamine and its complex shows that shift observed in C_7 and C_8 confirms the formation of complex and support the proposed structure.

3.6 Metal Analysis

Palladium(II) Sulfamethoxazole Complex

The percentage of the metal in the metal complexes is given in Table 3.9. The data shows that one molecule of the ligand is attached with palladium(II). The observed percentage of metal (Pd) nicely match with the calculated values for the formula $C_{10}H_{11}N_3O_3SCl_2Pd$.

Iron(II) Sulfamethoxazole Complex

Similar case is also observed for iron(II) sulfamethoxazole complex. One molecule of ligand is attached with iron(II). The percentage of metal determined by ICP-ES nicely match with the calculated values of the formula $C_{10}H_{11}N_3O_3SCl_2Fe$.

Zinc(II) Pyrimethamine Complex

Data in Table 3.9 shows that one molecule of ligand is attached with zinc(II). The percentage of metal in metal complex determined by ASS closely match with the calculated values for the formula $C_{12}H_{21}N_4O_4Cl_3Zn$.

The results obtained from ICP-ES, AAS and UV-visible for determining the percentage of metal in metal complexes confirms the proposed molecular weight.

Table 3.9: Percentage of metals in metal complexes.

Sr.No	Complex	Chemical formula	UV-Visible	AAS	ICP-ES	Theoretical
1	Palladium sulfamethoxazole	$C_{10}H_{11}N_3O_3SCl_2Pd$	22.25	–	23.9	24.7
2	Iron sulfamethoxazole	$C_{10}H_{11}N_3O_3SCl_2Fe$	12.9	–	14.5	14.7
3	Zinc pyrimethamine	$C_{12}H_{21}N_4O_4Cl_3Zn$	–	12.5	–	14.3

3.7 Antibacterial Activity

3.7.1 Sulfamethoxazole Complexes

Sulfamethoxazole is an antibiotic. It has a number of donor atoms (N, O, S) and thus can acts as a ligand.

It is a sulfa drug. Chiefly derived from sulfanilamide, chemically similar to PABA and capable of inhibiting bacterial growth and activity by interfering with the metabolic processes in bacteria that require PABA. The metabolism of sulfamethoxazole occurs predominantly by N-acetylation.

In this work, we have synthesized metal complexes of sulfamethoxazole with different metals such as palladium(II), iron(II), iridium(III) and tin(II). Antibacterial properties of these metal complexes have been checked against one gram positive (*Micrococcus luteus*) and five were gram negative (*Escherichia coli, Salmonella setubal, Pseudomonas picketii, Bordetella bronchiseptica* and *Enterobacter aerogens*) bacteria and results are summarized in Table 3.10. In order to compare the results obtained, the Roxyithromycin and Cefixime are used as standard drug. Biological activity of the complexes shows that compound iridium sulfamethoxazole and tin sulfamethoxazole shows the significant antibacterial activity as compared to standard drug against *Pseudomonas picketii, Bordetella bronchiseptica, Escherichia coli* and good antibacterial activity against *Micrococcus luteus*.

Table 3.10: Antibacterial activity of sulfamethoxazole complexes.

Sr.No	Ligand/ Complexes	Zone of Inhibition (mm)				
		P.picketti	*B.bronchiseptica*	*E.coli*	*M.luteus*	Activity
1	Palladium sulfamethoxazole	nil	nil	9	nil	Non significant
2	Iridium sulfamethoxazole	20	21	24	18	Significant
3	Tin sulfamethoxazole	25	25	23	15	Significant
4	DMSO	nil	nil	nil	nil	
5	Roxyithromycin	18	25	13	10	
6	Cefixime	nil	12	28	28	

3.7.2 Pyrimethamine Complexes

Biological activity of pyrimethamine shows that they show low antibacterial activity against *Bordetella bronchiseptica* and *Escherichia coli* as compared to standard drug Cefixime and Roxyithromycin. No activity is found for rest of bacterial strain as shown in Table 3.11.

Table 3.11: Antibacterial activity of pyrimethamine complexes.

Sr.No	Ligand/ Complexes	Zone Of Inhibition (mm)				
		P.picketti	*B.bronchiseptica*	*E.coli*	*M.luteus*	Activity
1	Zinc pyrimethamine	nil	17	9	10	Low
2	Cadmium pyrimethamine	nil	16	15	9	Low
3	DMSO	nil	nil	nil	nil	
4	Roxyithromycin	18	25	13	10	
5	Cefixime	nil	12	28	28	

REFERENCES

- Kesimili, B., Topacli,A., *spect.Acta*, **2001**, 57, p.1031
- Basolo, F., Johnson, R.C., "*Coordination Chemistry*", W.A.Benjamin, INC., New york, **1964**, p.1.
- Basolo, F., Johnson, R.C., "*Coordination Chemistry*", W.A.Benjamin, INC., New york, **1964**, p.131.
- Simplico, J., Wilkins, R.G., *J. Am. Chem. Soc.*, **1967**, 89, 6092.
- Spence, J.T., *Coord. Chem. Rev.*, **1969**, 4, 475.
- Fenery, R.E., Peterson, I.M., Sahinkeya, H., *J. Bacteriai*, **1957**, 73, 284.
- Pandey, K.K., Kaushal, R., *Indian J. Appl. Chem.*, **1969**, 32, p. 96.
- Chatturvedi, K.K., Kaushal, R., *Indian J. Pharm.*, **1975**, 37, p. 85.
- Kirschner, S., Wei, Y.K., Francis, D., Bergman, T.G., *J. Med Chem.* **1966**, 9, 369.
- Furst, A., "*The Chemistry of Chelation in cancer*", Thomas. Springfield, 3rd Ed., **1963**, 52.
- Lindoy, L.F., Livingstone, S.E., Nolam, J.D., (Unpublished results).
- Sawyer, A.K., Kuivila, H.G., *J. Org. Chem.*, **1962**, *27*, 610.
- Danish, M., Ali, S., Mazhar, M., *Main Group Met. Chem.*, **1996**, *19*, 121.
- Vatsa, C., Jain, V.K., Kesavadas, T., Tiekink, E.R.T., *J. Organomet. Chem.*, **1991**, *410*, 135.
- Anderson, H.H., *Inorg. Chem.*, **1964**, *3*, 912.
- Davies, A.G; Smith, L; Smith, P.J., *J. Organomet. Chem.*, **1972**, *39*, 279.

REFERENCES

- Ford, B.F.E; Liengme, B.V; Sams, J.R., *J. Organomet. Chem.*, **1969**, *19*, 53.
- Delmond, B; Pommier, J.C; Valade, J., *J. Organomet. Chem.*, **1973**, *47*, 339.
- Oclic, Y., Pereyre, M., *J. Organomet. Chem.*, **1973**, *55*, 273.
- Davies, A.G., Symes, W.R., *J. Chem. Soc.*, **1967**, 1009.
- Oclic, Y., Pereyre, M., *J. Organomet. Chem.*, **1973**, *55*, 273.
- Cohen, A.D., Dillard, C.R., *J. Organomet. Chem.*, **1970**, *25*, 421.
- Evan, C.J., Karpel, S., *"Organotin Compounds in Modern Technology", J. Organomet. Chem. Lib.*, Elsvier, Amsterdam, **1985**, 16.
- Van der Kerk, G.M., Noltes, J.G., Luijtien, J.G.A., *Angew. Chem.*, **1958**, *70*, 298.
- Plum, H., *Tin Its Uses,* **1981**, *127*, 7.
- Hobbs, L.A., Smith, P.J., *Tin Its Uses,* **1982**, *131,* 10.
- Tagliavini, G., *J. Organomet. Chem.*, **1992**, *437*, 15.
- Braunstein, A.E., *"Amino Group Transfer"* Boyer, in P.D. (ed.). "*The enzymes*", 3rd edn. Vol. 9, Pt. B, Acad. Press. N. Y. (1973) and *"Definite reviw of transammination reaction"*, **1973**, p.379.
- Lippard,J., "*Inorganic Chemistry in Biology and Medicine*", Edited by A. e Martell, (Acs, Sym, Ser. No. 140, Washington DC), **1980**.
- Patron,W.D.M., *Proc.Roy.Soc.*, **1961**, 154, 21.
- Brandt, W.W., Lwyer, F.P., Gyarfas, E.C., *Chem.Revs.*, **1954**, 54, 959.
- Forghieri, F., Preti, C., Tosi, G., P. Zonnini, P., *Aust. J. Chem.*, **1983**, 36, 1125.
- Kirchner, S., Wei, Y.K., Francis,D., Bergman, J.G., *J. Med. Chem.*, **1966**, 9, 369.

REFERENCES

- Sigel,H., McCromick, D.B., *Account Chem. Res.*, **1970**,3, 201.
- Bigotto, A., Costa, G., Mestroni, G., Pellizer, G., Puxeddu,A., Stefani, L., *Inorg. Chim. Acta. Rev.*, **1970**, 4, 41.
- Beleh, M.A., Medical Chemistry 412 Sulfonamide Lecture, University of Michigan, February, 4 & 6, **2003**.
- Hardman, J.C., Limbird, L.E., Gilamn, A., Goodman; Gilman, *The Pharmacological Basis of Therapeutics,* 10th Edition, New York, McGraw-Hill, **2001**, 1171.
- Williams, D.A., Lemke, T.L., *Foye's Principles of Medicinal Chemistry*, 5th Edition, Baltimore, M.D., Lippincott Williams & Wilkins, **2002**, 825.
- *Drug Information Handbook,* Hudson, O.H., Lexi-Comp., **2002**, 1100.
- *Drug Information Handbook,* Hudson, O.H., Lexi-Comp., **2002**, 1105.
- *Drug Information Handbook,* Hudson, O.H., Lexi-Comp., **2002**, 1104.
- Rudoy, R.C., Nelson, J.O., Haltalin, K.C., *Antimicrobial Agents and Chemothel.*, **1974**, 51439.
- Marrison, R.T., Boyd, R.N., *"Organic Chemistry"*, 6th Ed., **1996**, p.585.
- Dubdley, H., William Ian Fleming, "*Spectroscopic Methods in Organic Chemistry*", **1987**, p.40.
- Pavia, D.L; Lampman, G.M; Kriz, G.S., "*Introduction to Spectroscopy*", 3rd Ed, Thomson Learning, Inc., Australia, **2001**, p.26.
- Christian, G.D., "*Analytical Chemistry*", 5th Ed, John Wiley & Sons, Inc., New York, **1994**, p.465.
- Christian, G.D., "*Analytical Chemistry*", 5th Ed, John Wiley & Sons, Inc., New York, **1994**, p.467.
- Wruble,M., *J. Am. Pharm. Assoc.*, **1943**, 32, p. 80.
- Mellan,I., U.S. Patent, **1950**, 2,507,537.

REFERENCES

- Rosenzweig, S., Fuchs, W.M., U.S. Patent, **1951**, 2,536,095.
- Linter, C.J., U.S. Patent, **1957,** 2,776,970.
- Luckey,T.D., Venugopal, B., *Metal Toxicity in Mammals* **1**, Plenum Press, New York ,**1977**.
- Kanagaraj, G., Rao, G.N., *Synth. React. Inor. Met. Org. Chem.*, **1992**, 22 5, p. 559.
- Jigeesh, N., Murthy Krishna, P., Rangacharyulu, M., Premaswarup, D., *Indian J. Pure and Applied Physics*, **1989** 27, p. 197.
- Kesimili, B., Topacli,A., *J. Mol. Struct.,* **2001**, 561, p.55.
- Perrin, D.D., Armarego, W.L.F., Perrin, D.R., *"Purification of Laboraotry Chemicals",* Pergamon Press, Oxford, London, **1996**.
- Sandell, E.B., *"Colorimetric Determination of Trace Metal",* 3rd Ed., Interscience Publisher, Inc., N.Y., **1959**, p.711
- Vogel, A.I., *"A Textbook of Quantitative Chemical Analysis",* 5th Ed., Longman, London, **1997**, p.463.
- Ebsworth, E.A.V., Rankin, W.H., Gradock, S., *"Structural Methods in Organic Chemistry",* p.216.
- Pavia, D.L., Lampman, G.M., Kriz, G.S., *"Introduction to Spectroscopy",* 3rd Ed, Thomson Learning, Inc., Australia, **2001**, p.167.
- Pavia, D.L; Lampman, G.M; Kriz, G.S., *"Introduction to Spectroscopy",* 3rd Ed, Thomson Learning, Inc., Australia, **2001**, p. .

INDEX

V

Z

ABOUT AUTHOR

Iffat Imtiaz holds Master of Science in Chemistry and Master of Philosophy in Inorganic / Analytical Chemistry degrees from Quaid-i-Azam University, Pakistan. She has undergone training in the testing of drugs in Ministry of Health, Bahrain and taught chemistry at high school level. Moving ahead with great devotion and courage as she believes, "It's never too late to discover something new".

NOTES

NOTES

NOTES

NOTES

NOTES

www.ingramcontent.com/pod-product-compliance
Lightning Source LLC
LaVergne TN
LVHW090424210926
848480LV00002B/670
* 9 7 8 0 6 1 5 1 4 8 3 2 8 *